Analysis of Multiple Dependent Variables

POCKET GUIDES TO
SOCIAL WORK RESEARCH METHODS

Series Editor
Tony Tripodi, DSW
Professor Emeritus, Ohio State University

PATRICK DATTALO

Analysis of Multiple
Dependent Variables

OXFORD
UNIVERSITY PRESS

OXFORD
UNIVERSITY PRESS

Oxford University Press is a department of the University of Oxford.
It furthers the University's objective of excellence in research, scholarship,
and education by publishing worldwide.

Oxford   New York
Auckland   Cape Town   Dar es Salaam   Hong Kong   Karachi
Kuala Lumpur   Madrid   Melbourne   Mexico City   Nairobi
New Delhi   Shanghai   Taipei   Toronto

With offices in
Argentina   Austria   Brazil   Chile   Czech Republic   France   Greece
Guatemala   Hungary   Italy   Japan   Poland   Portugal   Singapore
South Korea   Switzerland   Thailand   Turkey   Ukraine   Vietnam

Oxford is a registered trademark of Oxford University Press in the UK and certain other
countries.

Published in the United States of America by
Oxford University Press
198 Madison Avenue, New York, NY 10016

Library of Congress Cataloging-in-Publication Data
Dattalo, Patrick.
Analysis of multiple dependent variables / Patrick Dattalo.
p. cm.—(Pocket guides to social work research methods)
Includes bibliographical references and index.
ISBN 978–0–19–977359–6 (pbk. : alk. paper)
1. Social service—Statistical methods.   2. Multivariate analysis.
I. Title.
HV29.D384 2013
519.5'35—dc23
2012029873

1 3 5 7 9 8 6 4 2
Printed in the United States of America
on acid-free paper

For Marie

Contents

Analysis of Multiple Dependent Variables

1

Basic Concepts and Assumptions

INTRODUCTION

Multivariate procedures allow social workers and other human services researchers to analyze multidimensional social problems and interventions in ways that minimize oversimplification. The term *multivariate* is used here to describe analyses in which several *dependent variables* **(DVs)** are being analyzed simultaneously. The term *univariate* is used to describe analyses in which one DV is being analyzed. DVs are also called criterion, response, *y*-variables, and endogenous variables. *Independent variables* **(IVs)** are also called predictor, explanatory, *x*-variables, and exogenous variables. For consistency, only the terms independent and dependent variables will be used here. Examples of multivariate analyses include investigations of relationships between personality traits and aspects of behavior, program characteristics and effectiveness, and client characteristics and service needs. The purposes of multivariate analyses include (1) data structure simplification and reduction, (2) description of relationships, and (3) prediction.

Using multivariate analysis for data structure simplification and reduction involves identifying and interpreting concepts called latent variables, or emergent variables. According to Huberty (1994), "what is being sought is an attribute or trait that may not be directly observable, and is not tangible; it is a latent variable often considered to comprise observable variables that are in some way combined (usually in a linear manner)" (p. 623).

Using multivariate analysis for description involves identifying and specifying the relative importance of IVs to anticipate the values of DVs; that is, description involves determining an ordering of IVs by answering the question: How well does the estimated relationship between a set of IVs and a set of DVs (i.e., a model) perform without each IV?

Using multivariate analysis for prediction involves identifying and specifying one or more IVs that can help anticipate the values of DVs. That is, the purpose of predictive analysis is to model the relationship between one or more IVs and one or more DVs, and to use this model to estimate values of the DVs when only values of the IVs are available.

Examples of multivariate statistical procedures to predict and describe relationships include *multivariate multiple regression* (MMR), *multivariate analysis of variance* (MANOVA), and *multivariate analysis of covariance* (MANCOVA). *Structural equation modeling* (SEM) may be used for data simplification and reduction, description, and prediction.

RATIONALE FOR MULTIVARIATE ANALYSIS

When multiple DVs are being modeled, a researcher faces a choice among whether to (1) analyze each DV separately; (2) aggregate the DVs before analysis (e.g., compute standard scores for each DV and add the scores of each DV to obtain a composite score for all DVs); or (3) perform a multivariate analysis, such as MANOVA. One approach that can help to clarify which of the aforementioned three choices is best is to examine the correlations among DVs. Low correlations (Pearson's $r < .20$) suggest the need to analyze each DV separately; moderate correlations ($r = .20$ to .50) suggest the need to analyze DVs together (i.e., a multivariate analysis); and high positive correlations ($r > .50$) suggest the need to aggregate the DVs before analysis (Stevens, 2009).

For example, MANOVA is conducted when the multiple DVs are conceptually or statistically related. When the DVs represent different dimensions of a multidimensional concept (e.g., clients satisfaction), they are conceptually related. DVs are statistically correlated when a measure of association, such as Pearson's, r, results...in a weak to moderate coefficient. Whether a researcher conducts a multivariate analysis or multiple univariate analyses the intercorrelations among DVs should be reported (Huberty & Morris, 1989; Tabachnick & Fidell, 2007). When there are moderate to low correlations among DVs, and a researcher wishes to retain separate measures of these DVs, an additional benefit is multiple operationalism. *Operationalism* refers to the process of representing constructs by a specific set of steps or operations. In other words, operationalism helps a researcher clarify the process of using empirical data to measure a concept (i.e., an abstract idea or theoretical construct), and to allow others to evaluate the validity of that measurement process. *Multiple Operationalism.* This term refers to the use of two or more measures to represent a concept. From this perspective, the use of multiple measures of a single concept provides a better chance of fully representing that concept. Campbell and Fiske (1959) explain that

"For any data taken from a single operation, there is a subinfinity of interpretations possible; a subinfinity of concepts, or combinations of concepts, that it could represent. Any single operation, as representative of concepts, is equivocal." (p. 101)

Boring (1953) stated, "...as long as a new construct has only the single operational definition that it received at birth, it is just a construct; when it gets two alternative operational definitions, it is beginning to be validated" (p. 183).

Multiple DVs may be suggested by theory, previously conducted empirical investigations, or practice experience. Moreover, as Campbell and Fiske (1959) have suggested, even if the primary interest is in only one DV, multiple operationalization may provide a more valid assessment of that DV. However, difficulty may arise in the interpretation of relationships among multiple DVs if there is reason to suspect that these variables are correlated. Consequently, the examination of correlations among DVs may provide a greater understanding than can be attained by considering these variables separately. The four multivariate procedures discussed in this book facilitate the analysis of correlated DVs.

Control of Type I Error. A second reason to conduct a multivariate analysis is to control type I error. Type I error is the probability of incorrectly identifying a statistically significant effect. Type I error is a false positive, and is the process of incorrectly rejecting the null hypothesis in favor of the alternative or research hypothesis. In an empirical investigation in which two or more comparisons are made from the data, there are two kinds of type I error: (1) the ***comparison-wise error rate*** is the probability of a type I error set by the researcher for evaluating each comparison, and (2) the ***study-wise error rate*** (sometimes termed ***experiment-wise error rate***) is the probability of making at least one type I error when performing a set of comparisons. The study-wise error rate (SW) is defined as $1 - (1 - \alpha)^k$, where α (alpha) is the significance level used for each test and k is equal to the number of comparisons. For example, if alpha is defined as $p < 0.05$, and tests are assumed to be independent, then, for a model with five predictor variables, SW $\leq 1 - (1 - 0.05)^5$, or $1 - 0.77 = 0.33$. Therefore, whenever multiple inferential statistical tests are performed, there is a potential problem of ***probability pyramiding***. Use of conventional levels of type I error probabilities for each test in a series of statistical tests may yield an unacceptably high type I error probability across all of the tests. If a researcher has a sound theoretical reason for performing sequential univariate or bivariate hypothesis tests, then he or she might consider an adjustment to the overall type I error probability. One approach is to use an ***additive inequality*** (i.e., Bonferroni). For m tests, the alpha level for each test (α_1) is given by the overall alpha level (α_m) divided by m; this function of α_1 should be considered as an upper bound for the overall alpha, α_m. Approaches to controlling type I error are discussed further in chapter 2.

Multivariate statistical procedures, however, are not panaceas; there are costs associated with their benefits. For, example, Bray and Maxwell (1985) state,

> ...the indiscriminant use of multiple dependent measures is never an appropriate substitute for a well-conceived study with a select number of variables...In fact, the use of too many dependent variables may hinder the chances of finding a significant result because of low statistical power or it may result in spurious findings due to chance. (pp. 9–10)

Following Enders (2003), the perspective taken here is to strongly encourage researchers to "think carefully when thinking multivariately" (p. 53). Accordingly, which procedure is most appropriate depends on the (1) purpose of the analysis; (2) sample size; and (3) tenability of assumptions.

These three issues will be discussed for MANOVA, MANCOVA, MMR, and SEM, which will be presented as alternative statistical procedures for analyzing models with more than one DV (see the chapter summaries below).

The remaining portion of this chapter is organized as follows: (1) audience and background; (2) level and type of examples; (3) definition of basic terms and concepts; (4) data assumptions; and (5) chapter summaries.

ORGANIZATION OF THE BOOK

Audience and Background

This book is intended primarily for producers and consumers of social work research, although it is expected that it will also be useful to other fields that use experiments and quasiexperiments, such as psychology, education, evaluation, medicine, and other health-related disciplines. Readers with a background in basic statistics are the primary audience. Ideally, readers will have taken a graduate-level statistics course with content that includes an introduction to ANOVA and regression modeling. However, each procedure will be presented here in sufficient detail to enable readers to implement that procedure without supplemental reading.

Level and Type of Examples

Discussion draws on the social work literature and on the literature from a variety of related disciplines. Note that all data and related files are posted on this book's companion website. Detailed social work–related examples are as follows:

1. Annotated examples of MANOVA and MANCOVA are provided using PASW. Computer output is presented as well as an explanation of how the output was produced. Instructions for using Stata and SAS to conduct MANOVA and MANCOVA also are provided;
2. An annotated example of MMR is provided using Stata. Computer output is presented as well as an explanation of how

the output was produced. Instructions for using PASW and SAS to conduct MMR also are provided; and

3. An annotated example of SEM is provided using AMOS. Computer output is presented as well as an explanation of how the output was produced.

Assumptions

The term *statistical model* is used here to describe mathematical statements about how variables are associated. Ideally, models are constructed to capture a pattern of association suggested by a theory. Models are constructed and *fitted* to statistically test theories that purport to explain relationships between and among variables. Statistical models include IVs and DVs. Variables also may be described in terms of order. *Higher-order variables* (e.g., *interactions* or *moderators*) are constructed from *lower-order variables* (i.e., *main effects*).

Model building concerns strategies for selecting an optimal set of IVs that explain the variance in one or more DVs (Schuster, 1998). A statistical model is said to be *misspecified* if it incorrectly includes unimportant variables or excludes important variables (Babyak, 2004). Including more variables than are needed is termed *overfitting*. Overfitting occurs when a model contains too many IVs in relation to sample size, and when IVs are intercorrelated. An overfitted model captures not only true regularities reflected in the data, but also chance patterns that reduce the model's predictive accuracy. Overfitting yields overly optimistic model results; that is, findings that appear in an overfitted model may not exist in the population from which a sample has been drawn. In contrast, *underfitting* occurs when a model has too few IVs. Underfitting results in omission bias, and, consequently, may have poor predictive ability because of the lack of detail in a regression model.

The simplest statistical model consists of IVs whose relationships with a DV are separate, and, consequently, there is *additivity*, or no interaction. A more complicated model is when the effect of one IV (X_1) on a DV (Y) depends on another IV (X_2). The relationship between X_1 and X_2 with Y is called a *statistical interaction* (or *moderation*). In a statistical test of interaction, the researcher explores this interaction in the regression model by multiplying X_1 by X_2, which produces X_1X_2, which is called an interaction of moderator term. Conversely, additivity means that the effect of one IV on the DV does not depend on the value of another IV. In a subsequent discussion, data being analyzed are assumed to follow the **general linear model (GLM).** One useful perspective on the GLM is as an extension of linear multiple regression, sometimes termed ordinary least squares regression. Multiple regression models a DV as a function of a set of IVs. This model seeks to explain a proportion of the variance in a DV at a statistically significant level, and can establish the relative predictive importance of the IVs.

The multiple regression equation for sample estimates takes the form

$$y = b_1 x_1 + b_2 x_2 + \cdots + b_n x_n + c \tag{1.1}$$

The b_i's are the regression coefficients, representing the amount the DV (y) changes when the IV (e.g., x_1) changes one unit. The c is the constant, or where the regression line intercepts the y-axis, and represents the value of the DV(y) when all IVs equal zero.

Ordinary least squares (OLS) regression derives its name from the method used to estimate the best-fit regression line: a line such that the sum of the squared deviations of the distances of all the points from the line is minimized. More specifically, the regression surface (i.e., a line in simple regression, or a plane or higher-dimensional surface in multiple regression) expresses the best predicted value of the DV (Y), given the values on the IVs (X's). However, phenomena of interest to social workers and other applied researchers are never perfectly predictable, and usually there is variation between observed values of a DV and those values predicted by OLS regression. The deviation of observed versus predicted values is called the **residual value.** Because the goal of regression analysis is to fit a linear function of the X variables as closely as possible to the observed Y variable, the residual values for the observed points can be used to devise a criterion for the "best fit." Specifically, in regression problems, the linear function is computed for which the sum-of-squared deviations of the observed points from values predicted by that function are minimized. Consequently, this general procedure is referred to as least squares estimation. *Power terms* (e.g., x^2, x^3) can be added IVs to explore curvilinear effects. *Cross-product terms* (e.g., $x_1 x_2$) can be added as IVs to explore interaction effects.

The GLM may be extended to include multiple DVs (i.e., multi-variate). This extension also allows multivariate tests of significance to be calculated when responses on multiple DVs are correlated. Separate uni-variate tests of significance for correlated DVs are not independent and may result in an erroneous decision about the statistical significance of predicted relationships. Multivariate tests of significance of linear combi-nations of multiple DVs also may provide insight into which dimensions of the DVs are, and are not, related to the IVs.

Important assumptions of the GLM, including (1) independence of observations, (2) model is specified correctly, (3) no missing data or data missing at random, (4) sound measurement, (5) orthogonality, (6) mul-tivariate normality, (7) absence of outliers, (8) linearity, and (9) homo-scedasticity (Stevens, 2009; Tabachnick & Fidell, 2007) are summarized below. Assumptions and remedies for the violation these assumptions are discussed further within the context of MANOVA, MANCOVA, MMR, and SEM.

Independence of Observations. This assumption is that the selection of any study participant is not related to the selection of any other partic-ipant. Kenny and Judd (1986) discuss three commonly assumed patterns of nonindependence: (1) nonindependence due to groups, (2) non-independence due to sequence, and (3) nonindependence due to space. *Nonindependence due to groups* can occur for a variety of reasons. For example, if members of the same group (e.g., students in a single school) are sampled, these students may be more similar in terms of measured and unmeasured characteristics. If characteristics are unmeasured, than the errors associated with their selection are not independent. If children are grouped in classrooms, group linkage by classrooms is likely because of a common teacher, classroom, or textbook. Social interaction may result in group linkage. In research on small groups, observations within groups are likely to be linked because the members of the groups interact during the course of the experimental session. *Nonindependence due to sequence* occurs when observations are repeatedly taken from a single unit over time. Such a data structure is commonly referred to as time-series data. *Nonindependence due to space* can occur when observations are made of units arranged in space (e.g., houses on a block, blocks within a city, or states within a country) may be spatially nonindependent. That is, observations that are nearer together in space may be more or less similar than those farther away in space.

According to Stevens (2009), violation of the independence of observations assumption is very serious. Only a small amount of dependence among observations has a dramatic effect on type I error. Specifically, when errors are positively correlated, nonindependence of within-group sample values increases type I error rates and at the same time decreases type II error probabilities. Moreover, the aforementioned situation does not improve with larger sample size; it becomes worse. Consequently, if there is reason to suspect dependence among observations, the researcher should consider using a more stringent level of significance (e.g., .01 rather than .05).

The independence of observations assumption fundamentally is a design issue (i.e., it is based on the way data are collected). Consequently, if nonindependence of observations is suspected, the most effective strategy is to use a research design that explicitly incorporates nonindependent (i.e., correlated) observations, such as multilevel modeling (Hox, 2002). For example, medical research applications often involve **hierarchical data structures** (i.e., nested) such as patients within hospitals or physicians within hospitals; for example, assessing differences in mortality rates across hospitals relative to a specific condition or procedure. Data are collected on random samples of patients nested within each hospital. In this application, it might be appropriate to adjust for covariates at both the patient-level (such as patient age, patient gender, and the severity of the index diagnosis) and at the hospital-level (such as hospital size and hospital teaching status). Hierarchical linear models, sometimes called multilevel linear models, nested models, mixed linear models, random-effects models, random parameter models, split-plot designs, or covariance components models handle parameters that vary at more than one level.

The basic logic behind these models is that a group mean (i.e., the potential **contextual effect**) can explain unique variance over and above an individual variable of the same name (Bickel, 2007). So, for instance, average group work hours may explain unique variance in individual well-being above individual reports of work hours. This occurs because there is no mathematical reason why the group-level relationship between means must be the same as the individual-level relationship. When the slope of the group-mean relationship differs from the slope of the individual-level relationship, then a contextual effect is present. To estimate contextual regression models, regression is used to simultaneously test the significance of the individual and group mean variables. If the group-mean variable is significant in this model it indicates that the individual-level and group-level slopes are significantly different, and one has evidence of a contextual effect.

Model Is Specified Correctly. As discussed above, this assumption concerns the accuracy of the statistical model being tested. Several tests of specification have been proposed. *Ramsey's Regression Specification Error Test (RESET)* is an F-test of differences of R^2 under linear versus nonlinear assumptions (Ramsey, 1969). More specifically, the RESET evaluates whether nonlinear combinations of the estimated values help explain the DV. The basic assumption of the RESET is that if nonlinear combinations of the explanatory variables have any power in explaining the endogenous variable, then the model is misspecified. That is, for a linear model that is properly specified, nonlinear transforms of the fitted values should not be useful in predicting the DV. While Stata, for example, labels the RESET as a test of omitted variables, it only tests if any nonlinear transforms of the specified DVs variables have been omitted. RESET does not test for other relevant linear or nonlinear variables that have been omitted.

The RESET (and other specification tests) should not be viewed as a substitute for a good literature review, which is critical to identify a model's variables. As a rule of thumb, the lower the overall effect (e.g., R^2 in multiple regression with a single DV), the more likely it is that important variables have been omitted from the model and that existing interpretations of the model will change when the model is correctly specified. Perhaps, specification error may be best identified when the research relies on model comparison as opposed to the testing of one model to assess the relative importance of the IVs.

Missing Data. Missing data in empirical social work research can substantially affect results. Common causes of missing data include participant nonresponse and research design. That is, some participants may decline to provide certain information, some data (e.g., archival) may not be available for all participants, and information may be purposely censored, for example, to protect confidentiality. There is a rich body of literature on missing data analysis. However, there is no consensus in the methodological literature about what constitutes excessive missingness (Enders, 2010; Little & Rubin, 2002). Based on whether missing data are dependent on observed values, patterns of missing data are classified into three categories: *missing completely at random* (MCAR), *missing at random* (MAR), and *missing not at random* (MNAR).

Missing completely at random (MCAR) exists when missing values are randomly distributed across all observations. Missing at random (MAR) is a condition that exists when missing values are not randomly distributed across all observations but are randomly distributed within one or more subsamples (i.e., missing more among whites than non-whites, but random within each subsample). When neither data are not MCAR or MAR, they are systematically missing or missing not at random (MNAR) (Heitjan, 1997; Kline, 2011). Biased parameter estimates may result in data that are systematically missing.

Missing data can be handled by either deletion or imputation techniques (Kline, 1998; Little & Rubin, 1987). Deletion techniques involve excluding participants with missing data from statistical calculations. *Imputation* is the process of replacing missing data with values that are based on the values of other variables in a data set. Unlike deletion, imputation retains sample size, and minimizes loss of statistical power. Imputation techniques are generally classified as either single or multiple. *Single imputation* assigns one estimate for each missing data point. Basically, *multiple imputation* is an extension of the single imputation idea, in which each missing value is replaced by a set of $m > 1$ plausible values to generate m apparently complete data sets. These m data sets are then analyzed by standard statistical software, and the results are combined using techniques suggested by Rubin (1987) to give parameter estimates and standard errors that take into account the uncertainty attributable to the missing data values. There is an increasing availability of software to perform multiple imputation, including PASW version 19. Free software to perform multiple imputation includes *NORM*, which is available as a standalone or an S-Plus version. Version 2.03 of NORM can be downloaded from http://sites.stat.psu.edu/~jls/norm203.exe

Sound Measurement. Data analysis is influenced at least in part by how and how well concepts have been operationalized. A thorough discussion of measurement theory and psychometric technique is beyond the scope of this text. However, one important principle related to *how* concepts have been operationalized is level of measurement. Two important principles related to *how well* concepts have been operationalized are reliability and validity.

Levels of measurement are expressions that refer to the **theory of scale types** developed by the psychologist Stanley Smith Stevens. Stevens (1946, 1951) argued that all empirical measures may be categorized as nominal, ordinal, interval, or ratio, from lowest to highest levels of measurement. A variable may be defined as an operationalized concept that changes (e.g., gender) and therefore is a logical grouping of alternative empirical values (e.g., female or male). The level of measurement, then, refers to the relationship among these possible alternative empirical values. Stevens argued that an understanding of a variable's level of measurement helps to clarify how a variable changes and therefore, elucidates how a concept is empirically measured. Knowing the level of measurement of a variable helps to identify which statistical procedure is most appropriate, since the level of measurement determines the level of mathematical precision.

Stevens's theory of scale types posits a hierarchy of levels of measurement. The GLM assumes that variables are operationalized at the highest practical level of measurement. At lower levels of measurement, assumptions tend to be less restrictive and data analyses tend to be less sensitive. As one moves from lower to higher levels in the hierarchy, the current level includes all of the characteristics of the one below it, plus something new. At the **nominal level**, alternative empirical values are mutually exclusive categories; no ordering of alternative values (e.g., more versus less) is implied. For example, gender is a nominal level variable. At the **ordinal level** alternative empirical values may be ranked. For example, on a survey you might code educational attainment as 0 = less than high school, 1 = some high school, 2 = high school degree; 3 = some college; 4 = college degree; 5 = post college. In this measure, higher numbers mean more education. But, the distance between values is not assumed to be equal. At the interval level the distance between values is assumed to be equal. For example, on the Fahrenheit temperature scale, the distance from 30 to 40 is same as distance from 70 to 80. The interval between values is interpretable, and, consequently, the average of a group of values on an interval variable may be computed. At the ratio level of measurement there is an absolute zero. This means that you can construct a meaningful fraction (or ratio) with a ratio variable. Weight is a ratio variable. It should be noted that although Stevens's classification is widely adopted, it is not universally accepted (cf. Velleman & Wilkinson, 1993; Chrisman, 1998).

Reliability refers to consistency. A reliable measure is one that yields consistent results (given an unchanged measured phenomenon) over repeated measurements. *Validity* refers to accuracy. A valid measure is one that measures what it purports to measure. Measurement reliability is necessary, but not sufficient, for measurement validity. Accordingly, for example, if scale does not accurately measure what it is supposed to measure, there is no reason to use it, even if it measures consistently (reliably). It is expected that measures are sufficiently valid, and reliability is a logical first step for demonstrating validity (i.e., to be accurate a measure must be consistent).

A valid measure (e.g., scale), which by definition is a reliable measure, should yield an outcome (e.g., score) that has minimum measurement error. According to classical measurement theory, an observed score is composed of (1) a true value and (2) error. It is assumed that measurement error is minimal. *Measurement error* is either systematic or random. As the term implies, *random error* differentially affects members of a sample. For instance, study participants' moods can inflate or deflate their performance. Because the sum of all random errors in a distribution equals zero, random error adds variability to data, but do not affect average performance. Consequently, random error is sometimes referred to as noise. In contrast, *systematic error* tends to affect all members of a sample. Consequently, systematic measurement error is problematic. Systematic error can be detected by assessing reliability and validity.

A detailed discussion of strategies to assess measurement reliability and validity is beyond the scope of this chapter, but commonly used approaches to assessing reliability include internal consistency, test–retest, and parallel forms. Commonly used approaches to assessing validity include face, content, criterion, construct, and factorial. See Rubin and Babbie (2010) for a detailed discussion of strategies to assess reliability and validity.

Commonly used strategies to minimize both random and systematic measurement error include: (1) pilot testing of measurement instruments by collecting feedback from respondents about clarity in terms of question organization and meaning; (2) ensuring consistency in how measurement instruments are used by training proctors, interviewers, and observers; (3) verifying data are inputted accurately for computer analysis; and (4) using multiple measures of the same concept to triangulate responses.

Absence of Multicollinearity is when variables are highly correlated, and *singularity* is when variables are perfectly correlated. As a rule of thumb, intercorrelation among IVs above .80 signals a possible problem. Likewise, high multicollinearity is signaled when high R^2 and significant F-tests of the model occur in combination with nonsignificant t-tests of coefficients. Large standard errors because of multicollinearity result in a reduced probability of rejecting the null hypothesis (i.e., power) and wide confidence intervals. Under multicollinearity, estimates are unbiased, but assessments of the relative strength of the explanatory variables and their joint effect are unreliable. That is, under multicollinearity, beta weights and R^2 cannot be interpreted reliably even though predicted values are still the best estimate using the given IVs.

Orthogonality. Perfect no association between IVs is preferred so that each IV adds to the ability of the set of IVs to predict values of the DV. Consequently, if the relationship between each IV and each DV is orthogonal, the effect of an individual IV may be isolated.

Commonly used approaches to identifying multicollinearity include (1) inspection of bivariate correlations among IVS; (2) calculating *tolerance*, which is defined as $1 - R^2$, where R^2 is the multiple R of a given IV regressed on all other IVs. If the tolerance value is less than some threshold, usually .20, multicollinearity is suspected; and (3) calculating the *variance inflation factor* (**VIF**), which is the reciprocal of tolerance. The rule of thumb is that VIF > 4.0 (some researchers prefer VIF > 5.0) when multicollinearity is a problem.

Multivariate Normality. Univariate normality is a necessary, but not sufficient, condition for multivariate normality to exist (Stevens, 2009). Multivariate normality exists when each variable has a normal distribution about fixed values on all other variables. Two other properties of a multivariate normal distribution are as follows: (1) any linear combinations of the variables are normally distributed; and (2) all subsets of the set of variables have multivariate normal distributions. This latter property implies that all pairs of variables must be bivariate normal.

A multivariate normal distribution has zero skew and zero kurtosis. *Skew* is the tilt (or lack of it) in a distribution. That is, residuals are not symmetrically distributed about the mean error. Negative skew is right-leaning and positive skew is left-leaning. An S-shaped pattern of deviations indicates that the residuals have excessive kurtosis. *Kurtosis* is the peakedness of a distribution. *Leptokurtosis* is a distribution with "fat tails," and there is a relatively small proportion of values both above and below the mean residual. *Platykurtosis* is a distribution with "thin tails," and there are a relatively large proportion of values both above and below the mean residual.

According to Stevens (2009), deviation from multivariate normality has only a small effect on type I error. Multivariate skewness appears to have a negligible effect on power. However, Olsen (1974) found that platykurtosis does have an effect on power, and the severity of the effect increases as platykurtosis spreads from one to all groups.

Several strategies to assess multivariate normality are as follows:

1. Bivariate normality, for correlated variables, implies that the scatterplots for each pair of variables will be elliptical; the higher the correlation, the thinner the ellipse. Therefore, as a partial check on multivariate normality, scatterplots for each pair of variables could inspected for the hoped for approximately elliptical pattern;

2. Another approach is to focus on the distribution of a model's error or residuals. For a data set, residual values are the differences between values predicted by a model and observed values of data. To assess the normality of residuals, predicted values of a DV are plotted on the y-axis of a graph and residual values of this same DV are plotted on the x-axis this graph. Under perfect normality, the plot will be a 45-degree line. Univariate normality is necessary, but not sufficient, for multivariate normality. However, the examination of residuals may allow a researcher to assess both the normality of residuals, and the more important homoscedasticity assumption discussed below;

3. Looney (1995) suggests using battery of tests of univariate normality. That is, as a first step in assessing multivariate normality is to test each variable separately for univariate normality. Of the many procedures available for assessing univariate normality, two of the most commonly used are (1) an examination of skewness and kurtosis, and (2) the Shapiro–Wilk (1965) W test. Many authors have recommended that both approaches be used, as neither is uniformly superior to the other in detecting nonunivariate alternatives. If either test strongly indicates a departure from normality, then normality is rejected; and

4. DeCarlo (1997) recommends that, as a first step in assessing multivariate normality, each variable be separately tested for univariate normality. Because several tests are performed, a **Bonferroni correction** can be used to control the type I error rate (discussed in detail in chapter 2). Although univariate normality is a necessary condition for multivariate normality, it is not sufficient, which means that a nonnormal multivariate distribution can have normal marginals. Therefore, if univariate normality is not rejected, then the next step is to check for multivariate normality. DeCarlo (1997) provides a macro that can be used to assess both univariate and multivariate normality. For univariate data, the macro calculates the **kurtosis statistic b_2**, and the **skewness statistic $\sqrt{b_1}$**. It also provides two omnibus tests: K^2, which sums the two chi-squares for skewness and kurtosis, and a **score (Lagrange multiplier) test**, which is a function of $\sqrt{b_1}$ and b_2.

   In addition to the univariate statistics, the macro provides the following measures of multivariate normality: (a) Mardia's (1970) multivariate kurtosis; (b) Srivistava's (1984) and Small's (1980) tests of multivariate kurtosis and skew, both of which are discussed by Looney (1995); (c) an omnibus test of multivariate normality based on Small's statistics (see Looney, 1995); (d) a list of the five cases with the largest squared **Mahalanobis distances**; (e) a plot of the squared Mahalanobis distances, which is useful for checking multivariate normality and for detecting multivariate outliers; and (f) Bonferroni adjusted critical values for testing for a single multivariate outlier by using the Mahalanobis distance, as discussed by Penny (1996). This macro is demonstrated in the annotated example of MANOVA in chapter 2.

   Absence of Outliers. Unusual or extreme values are defined as observations that appear to be inconsistent with other observations in the data set. Outliers can occur by chance in any distribution. In the case of normally distributed data, roughly 1 in 22 observations will differ by twice the standard deviation or more from the mean, and 1 in 370 will deviate by three times the standard deviation. Outliers, however, may result from data entry error. However, outliers also may be the result of incorrect distributional assumptions (e.g., treating count data as normally distributed data).

MANOVA is sensitive to the effect of outliers, which may affect the type I error rate; OLS regression analysis also is sensitive to outliers. Outliers in OLS regression can overstate the coefficient of determination (R^2), and give erroneous values for the slope and intercept. Within the context of the assumption of the absence of outliers, there are three important characteristics of potentially errant or seemingly unusual observations (data points). The first is *leverage*. In regression, high leverage cases are those with unusual values on the IVs. In MANOVA, high leverage cases are those with an unusual combination of scores on the DVs. The second is *discrepancy (or distance^2)* between predicted and the observed values on the DV. The third is *influence*, which reflects the product of leverage and discrepancy.

Lorenz (1987) suggests using **Cook's D** as the focus of a procedure for pre-screening for outliers. Cook's distance measure D provides an overall measure of the impact of an observation on the estimated regression coefficient. Just because the residual plot or Cook's distance measure test identifies an observation as an outlier does not mean one should automatically eliminate the point. Some possible approaches to working with outliers are as follows:

1. Transformation of data is one way to reduce the impact of outliers because the most commonly used expressions, square roots and logarithms, shrink larger values to a much greater extent than they shrink smaller values. However, transformations may not fit into the theory of the model or they may affect its interpretation. Taking the log of a variable does more than make a distribution less skewed; it changes the relationship between the original variable and the other variables in the model. In addition, most commonly used transformations require non-negative data or data that are greater than zero, so they are not always the answer;

2. Deletion of outliers. When in doubt, report model results with and without outliers; and

3. Use of methods that are robust in the presence of outliers, such as robust regression models.

At a minimum, researchers should consider fitting a model with and without outliers, and compare the coefficients, mean-squared error and, R^2 from the two models. It also may be useful for the researcher to conjecture about possible reasons for outliers. Conjecture may facilitate conclusions about the representativeness of a sample with outliers. Conjecture also may assist in the design of future studies by suggesting possible strategies to prevent missing data.

Linearity. This is the assumption that there is a straight line relationship between variables. Violations of linearity are of concern because if a linear model is fitted to data that are nonlinearly related, predictions are likely to have large residuals. Nonlinearity is usually most evident in a plot of the observed versus predicted values, or a plot of residuals versus predicted values. Points should be symmetrically distributed around a diagonal line in the former plot, or a horizontal line in the latter plot.

When nonlinearity is present, one solution is to apply a *nonlinear transformation* to the IVs or DVs. For example, if the data are strictly positive, a log transformation may be feasible. Another possibility is to consider adding another IV that is a nonlinear function of one of the IVs (e.g., X^2).

Homoscedasticity. The assumption that the variability in scores for one variable is equal at all values of another variable is termed homoscedasticity. Violations of homoscedasticity make it difficult to estimate standard error, and usually results in confidence intervals that are too wide or too narrow. That is, heteroscedasticity can result in biased parameter estimates, and standard error estimates that are either too large or too small. Consequently, heteroscedasticity can increase type I error, or increase type II error. One approach to detecting violations of homoscedasticity is to examine plots of residuals versus predicted values for evidence that residuals are a function of predicted values. However, heteroscedasticity also can be a result of a violation of the linearity or normality assumptions, and may also be remedied by addressing those violations.

It is important to note that rarely will a strategy satisfy all assumptions of the GLM. The flexibility of the general linear model allows it to be applied to a variety of research designs. When available, information is presented about the performance of a strategy when the assumptions of the GLM are not tenable.

CHAPTER SUMMARIES

MANOVA, MANCOVA, MMR, and SEM are discussed in terms of (1) purpose of the analysis; (2) important assumptions; (3) key concepts and analytical steps; (4) sample size and power requirements; (5) strengths and limitations; (6) an annotated example; (7) reporting the results of an analysis; and (8) additional examples from the applied research literature.

Chapter 2—Multivariate Analysis of Variance

MANOVA is discussed. In analysis of variance (ANOVA), the mean differences between three or more groups on a single DV are evaluated. Social workers and other applied researchers may also want to evaluate the mean differences between groups on two or more DVs. That is, in MANOVA, there is at least one IV (i.e., group) with two or more levels (i.e., subgroups), and at least two DVs. Whether groups differ significantly on one or more optimally weighted linear combinations of two or more DVs may be evaluated. Some researchers argue that MANOVA is preferable to performing a series of ANOVAs (one for each DV) because (1) multiple ANOVAs can capitalize on chance and (2) ANOVA ignores intercorrelations among DVs. In contrast, MANOVA controls for intercorrelations among DVs.

Chapter 3—Multivariate Analysis of Covariance

MANCOVA is similar to MANOVA, but allows a researcher to control for the effects of supplementary continuous IVs, termed covariates. In an experimental design, covariates are usually the variables not controlled by the experimenter, but still affect the DVs.

Chapter 4—Multivariate Multiple Regression

MMR, which models the relationship between several DVs and a set of IVs, is presented. The DVs may be the multiple dimensions of a single concept. For example, one might conceptualize consumer satisfaction as consisting of three dimensions: (1) congruence between provider and consumer service delivery expectations; (2) provider responsiveness, including courtesy and timeliness; and (3) effectiveness of service interventions. MMR estimates the same coefficients and standard errors as one would obtain using separate OLS regressions (one for each criterion variable). In addition, MMR is a joint estimator and also estimates between-equation covariances. This means that it is possible to test coefficients (IVs) across equations (DVs).

That is, although MMR's multivariate tests (multivariate F-tests) take into account the correlations among the criterion variables, the regression coefficients and univariate F-values identified by MMR will be exactly the same as those obtained from performing one regression analysis for each DV. For q DVs, q regression formulas may be estimated. The utility of MMR lies in its ability to test significance across and between the q equations. At least three kinds of statistical significance questions may be asked. First, is there a relationship between the DVs taken as a set and the IVs taken as a set? Second, is a particular IV or subset of the IVs related to the set of DVs? Third, is the relationship between a set of IVs and a particular DV statistically significant, when controlling for the association among the set of DVs?

Chapter 5—Structural Equation Models

SEM, also termed analysis of covariance structures, is discussed. SEM refers to a hybrid model that integrates path analysis and factor analysis. SEM is used when data consist of multiple indicators for each variable (called latent variables or factors) and specified paths connecting the latent variables. Thinking of SEM as a combination of factor analysis (to be discussed briefly in chapter 4) and path analysis (also to be discussed briefly in chapter 4) ensures consideration of SEM's two primary components: the measurement model and the structural model. The measurement model describes the relationships between observed variables and the construct or constructs those variables are hypothesized to measure. The structural model describes interrelationships among constructs.

When the measurement model and the structural model are considered together, the model is termed the composite or full structural model. In SEM, statistical power is the ability to detect and reject a poor model. In contrast to traditional hypothesis testing, the goal in SEM analysis is to produce a nonsignificant result (i.e., to fail to reject the null of no difference between the proposed and the perfect model). The null hypothesis is assessed by forming a discrepancy function between the proposed model (specified by the researcher) and the perfect or saturated model (one with no constraints that will always fit any data perfectly).

Various discrepancy functions can be formed depending on the particular minimization algorithm being used (e.g., maximum likelihood), but the goal remains the same: to derive a test statistic that has a known distribution, and then to compare the obtained value of the test statistic against tabled values in order to make a decision about the null hypothesis. Because in SEM the researcher is attempting to develop a theoretical model that accounts for all the covariances among the measured items, a nonsignificant difference between the proposed model and the saturated model is argued to be suggestive of support for the proposed model.

Chapter 6—Choosing among Procedures for the Analysis of Multiple Dependent Variables

Conclusions and recommendations are summarized, and MANOVA, MANCOVA, MMR, and SEM are compared and contrasted.

2

Multivariate Analysis of Variance: Overview and Key Concepts

This chapter presents MANOVA as a generalization of the *t*-test and ANOVA. The **t-test** is a strategy to test hypotheses about differences between two groups on a single mean. When there are more than two means, it is possible to use a series of *t*-tests to evaluate hypotheses about the difference between each pair of group means (e.g., for three groups, there are three unique pairs of means). However, conducting multiple *t*-tests can lead to inflation of the type I error rate (see chapter 1 for a discussion of type I and type II errors). Consequently,

1. **ANOVA** is used to test hypotheses about differences between three or more groups on a single mean; and
2. **MANOVA** is a strategy to test hypotheses about differences between two or more groups on two or more means (i.e., a vector of means).

As a generalization of the *t*-test and ANOVA, MANOVA may be understood as the ratio between two measures of multivariate variance. **Univariate variance** is defined as the average squared deviation of each value of a variable from that variable's mean.

Multivariate variance is defined as the simultaneous average squared deviation of each value of a variable on multiple means. Univariate variance and multivariate variance will be discussed further in the ANOVA and MANOVA sections below.

THE t-TEST

Frequently, hypotheses concern differences between the means of two groups in a sample. For example, a hypothesis could state that, in a random sample, the average age of males does not equal the average age of females. The t-test may be used to test the significance of this difference in average age between males and females. The variances of the two sample means may be assumed to be equal or unequal. There are two basic designs for comparing the mean of one group on a particular variable with the mean of another group on that same variable. Two samples are **independent** if the data in one sample are unrelated to the data in the other sample. *Two samples are **paired** (also termed **correlated** or **dependent** samples)* if each data point in one sample is matched to a unique data point in the second sample. An example of a paired sample is a pre-test/post-test study design in which all participants are measured on a DV before and after an intervention.

William Gosset, who published under the pseudonym of Student, noted that using a sample's standard deviation to estimate the population's standard deviation is unreliable for small samples (Student, 1908). This unreliability is because the sample standard deviation tends to underestimate the population standard deviation. As a result, Gosset described a distribution that permits the testing of hypotheses from normally distributed populations when the population mean is not known. This distribution is the **t-distribution** *or* **Student's t.**

The t-test is used when a population's variance is not known and the sample's size is small ($N < 30$). In fact, even when $N > 30$, the t-distribution is more accurate than the normal distribution for assessing probabilities, and, therefore, t is the distribution of choice in studies that rely on samples to compare two means.

The t-distribution is similar to the normal distribution when the estimate of variance is based on many *degrees of freedom (df)* (i.e., larger samples), but has relatively more scores in its tails when there are fewer

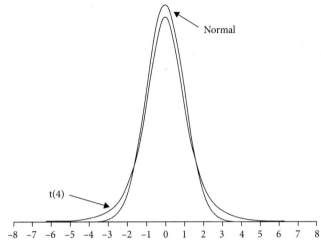

Figure 2.1 *t* versus Normal Distribution.

degrees of freedom (i.e., smaller samples). The *t*-distribution has $n - 1$ degrees of freedom, where n is the sample size. Figure 2.1 shows the *t* distribution with four *df* and the standard normal distribution. Note that the normal distribution has relatively more scores in the center of the distribution and the *t*-distribution has relatively more in the tails. The *t*-distribution, therefore, is leptokurtic. Because the *t*-distribution is leptokurtic (i.e., flatter), the percentage of the distribution, for example, within 1.96 standard deviations of the mean, is less than the 95% for the normal distribution.

Important properties of the *t*-distribution are as follows: (1) different for different sample sizes; (2) generally bell-shaped, but with smaller sample sizes shows increased variability; (3) as the sample size increases, approaches a normal distribution; (4) symmetrical about the mean; (5) the mean of the distribution is equal to 0; and (6) variance is equal to $v / (v - 2)$, where v is the degrees of freedom and $v \geq 2$.

The *t*-statistic is similar to the **z-statistic**: both statistics are expressed as the deviation of a sample mean from a population mean in terms of a standardized measure of distance (i.e., the standard error of the mean and the standard deviation, respectively). The *t*-test evaluates this difference in terms of the probability of obtaining an observed difference as large as or larger than the sample difference, if the null hypothesis of no difference is true. The α-level (e.g., .05) defines "likely." That is, if α is set at $p = .05$, then unlikely is .05 or less. If the α of an observed difference is equal to or less than .05, then the observed difference is unlikely if the null hypothesis is true. As a result, the null hypothesis is treated as false and rejected. In contrast, if the α of the observed difference is greater than .05, then the null is retained.

The following procedure is used to test whether a difference between two means in a sample is likely to be true at some level of probability (e.g., .05) for the population:

1. Formulate a hypothesis (variously referred to as the research, alternative, or H_1) about the difference;
2. State what would be true if H_0 is false (i.e., the null hypothesis or H_0);
3. Measure the difference in a random sample drawn from a population;
4. Assuming either that the variances around the means of the two groups being compared are equal or are unequal, evaluate the probability of obtaining the sample results if the H_0 is true; and
5. If these results are unlikely, reject the H_0.

ANOVA

When there are more than two means, conducting multiple t-tests can lead to *inflation of the type I error rate*. This refers to the fact that the more comparisons that are conducted at, for example, $\alpha = .05$, the more likely that a statistically significant comparison will be identified. For example, if the comparisons are independent, with five means there are 10 possible *pair-wise comparisons*. Doing all possible pair-wise comparisons on five means would increase the overall chance of a type I error to

$$\alpha^* = 1 - (1 - \alpha)^{10} = 1 - .599 = .401 \qquad (2.1)$$

For five comparisons at $\alpha = .05$ there is a 40.1% chance of making a type I error, instead of 5%. The overall chance of a type I error rate in a particular series of pair-wise comparisons is referred to as the *family-wise error rate*.

Introduced by Fisher (1918), ANOVA is used to test hypotheses about differences between two or more means without increasing the type I error rate. ANOVA may be defined as the ratio of two univariate variances: (1) **sum-of-squares-between** (or **SSB**) is a measure of the variability of the values of a variable (e.g., age) around the grand mean (i.e., the mean of several groups) of the variable; and (2) **sum-of-squares-within**, also termed **SSW**, **residual sum-of-squares**, or **sum-of-squares error**, is a measure of the variability within respective groups around that group's mean for the same variable (e.g., age). That is, the total variation (**sum-of-squares-total or SST**) of the observations is **partitioned** (or **divided**) into a part attributable to differences between group means and a part attributable to the differences between observations in the same group, or SST = SSB + SSW, and SSB/SSW, equals F (also termed **F-ratio** and **F-test**), which is used to assess the variability of the group means.

The shape of the distribution of the values of the F distribution depends on the sample size. More precisely, it depends on two degrees of freedom parameters: one for the numerator (SSB) and one for the denominator (MSW). SSB has $k - 1$ degrees of freedom, and MSW has $N - k$ degrees of freedom, where N is the total number of observations for the groups, and k is the number of groups. To summarize:

$$df_{SSB} = k - 1 \qquad (2.2)$$

$$df_{SSW} = N - k \qquad (2.3)$$

and

$$F = SSB/SSW \qquad (2.4)$$

If the means of the populations represented by the groups are equal, then, within the limits of random variation, the SSB and SSW should be equal; that is, $F = 1$. Deviations from 1 are assessed by comparing the observed F to the distribution of values of F if the null hypothesis that group means are equal is true.

More specifically, the larger the value of F, the greater the evidence that group means differ. The observed value of F is compared with a table of values of the F-distribution. It should be noted that, with two groups, one-way analysis of variance is equivalent to the two-sample t-test, and $F = t^2$.

Necessary assumptions for the F-test include the following: (1) the response variable is normally distributed in each group; and (2) the groups have equal variances. Note that the aforementioned assumptions are parametric assumptions and that ANOVA (as illustrated here) also requires that the observations are independent. F is an extended family of distributions, which varies as a function of a pair of df (one for SSB and one for SSW). F is positively skewed. Figure 2.2 illustrates an F distribution with degrees of freedom $(df) = df_1, df_2$, or $df = 5,10$. F ratios, like the variance estimates from which they are derived, cannot have a value less than zero. Consequently, in one sense, F is a one-tailed probability test. However, the F-ratio is sensitive to any pattern of differences among means. Therefore, F should be more appropriately treated as a two-tailed test.

When the overall F-test is not significant (i.e., assuming the hull hypothesis is true, the observed value of F is likely or greater than a pre-established α-level, such as $p = .05$, analysis usually terminates. When differences are statistically significant (i.e., assuming the hull hypothesis is true, the observed value of F is unlikely or less than a pre-established α-level, such as $p = .05$), the researcher will often identify specific differences among groups with post hoc tests. That is, with ANOVA, if the null hypothesis is rejected, then at least two groups are different in terms of the mean for the group on the DV. To determine which groups are different, *post hoc tests* are performed using some form of correction. Commonly used post hoc tests include ***Bonferonni***, ***Tukey b***, ***Tukey***, and ***Scheffé***. Post hoc tests are discussed further within the context of MANOVA.

In general, post hoc tests will be robust in those situations where the one-way ANOVA's F-test is robust, and will be subject to the same potential problems with unequal variances, particularly when the sample sizes are unequal.

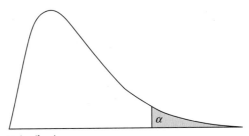

Figure 2.2 The F Distribution

MANOVA

Developed as a theoretical construct by Wilks (1932), MANOVA designs are substantially more complicated than ANOVA designs, and, therefore, there can be ambiguity about the relationship of each IV with each DV. In MANOVA, there is at least one IV (i.e., group) with two or more levels (i.e., subgroups), and at least two DVs. MANOVA designs evaluate whether groups differ on at least one optimally weighted linear combination (i.e., *composite means* or *centroids*) of at least two DVs.

MANOVA, therefore, is an ANOVA with several DVs. The testing of multiple DVs is accomplished by creating new DVs that maximize group differences. The gain in power obtained from decreased within-group sum of squares might be offset by the loss in degrees of freedom. One degree of freedom is lost for each DV. Some researchers argue that MANOVA is preferable to performing a series of ANOVAs (i.e., one for each DV) because (1) multiple ANOVAs can capitalize on chance (i.e., inflate α or increase type I error); and (2) ANOVA ignores intercorrelations among DVs; in contrast, MANOVA controls for intercorrelations among DVs. See chapter 1 for additional discussion of intercorrelations among DVs and inflated α–levels.

A *two-group MANOVA*-termed *Hotelling's T^2* is used when one IV has two groups and there are several DVs (Hotelling, 1931). For example, there might be two DVs, such as score on an academic achievement test and attention span in the classroom, and two levels of types of educational therapy, emphasis on perceptual training versus emphasis on academic training. It is inappropriate to use separate t tests for each DV to evaluate differences between groups because this strategy inflates type I error. Instead, Hotelling's T^2 is used to see if groups differ on the two DVs combined. The researcher asks if there are non-chance differences in the centroids (average on the combined DVs) for the two groups. Hotelling's T^2 is a special case of MANOVA, just as the t-test is a special case of ANOVA, when the IV has only two groups.

One-way MANOVA *evaluates* differences among optimally weighted linear combinations for a set of DVs when there are more than two level levels (i.e., groups) of one IV (factor). Any number of DVs may be used; the procedure addresses correlations among them, and controls for type I error. Once statistically significant differences are found, planned and post hoc comparisons (discussed below) are available to assess which DVs are influenced by the IV.

Factorial MANOVA is an extension of MANOVA to designs with more than one IV and multiple DVs. For example, a researcher may wish to explore differences in academic achievement (DV) and attention span (DV) based on gender (a between-subjects IV) and educational therapy (another between-subjects IV). In this case, the analysis is a *two-way between-subjects factorial MANOVA* that tests the main effects of gender and type of educational therapy and their interaction on the centroids of the DVs.

As stated earlier, ANOVA may be defined as the ratio of two univariate variances. Analogously, MANOVA may be defined as the ratio of two multivariate variances; *multivariate variance* is a measure of the simultaneous dispersion of values around multiple means. More specifically, MANOVA may be described as the ratio of the *determinants* of two *variance–covariance matrices*. An understanding of this definition requires an understanding of three key concepts: (1) matrix, (2) determinant of a matrix, and (3) variance–covariance matrix. For readers interested in a more detailed discussion of these concepts, an introduction to matrix algebra is posted on this book's website.

1. An *m* × *n* **matrix** is a rectangular array of real numbers with *m* rows and *n* columns. Rows are horizontal and columns are vertical;

2. One measure of multivariate variance is the **determinant of a matrix**. The 2×2 matrix $A = \begin{bmatrix} A & B \\ C & D \end{bmatrix}$ has a determinant, defined as $|A| = ad - bc$. A *determinant* is a real number associated with every square matrix. Mathematically, therefore, the determinant is the product of the elements on the main diagonal minus the product of the elements off the main diagonal, and it: (1) is a real number; (2) can be a negative number; and (3) only exists for square matrices. An important rule about determinants is that $|A|$ is always 0 if, and only if, at least two vectors of a matrix are linearly dependent. A vector is defined as a row (horizontal) or a column (vertical) of numbers. In general, the determinant increases with increasing independence of the vectors constituting the matrix.

The determinant can be conceptualized as the area (two-dimensional) or volume (three- or more dimensional) of a geometrical shape, which is spanned by the vectors of a matrix (e.g., a parallelogram with sides formed by two row and two column vectors). A parallelogram is a quadrilateral that has two pairs of parallel sides. The area of the parallelogram is the absolute value of the determinant of the matrix formed by the vectors representing the parallelogram's sides. The more similar the vectors (i.e., they point in the same direction), the smaller the area or volume and the smaller the determinant. In higher dimensions, the analog of volume is called hypervolume and the same conclusion can be drawn by the same argument: the hypervolume of the parallel-sided region determined by k vectors in k dimensions is the absolute value of the determinant whose entries are their components in the directions of the these vectors.

For example, in Figure 2.3, parallelogram 1 has a greater volume (e.g., determinant) than parallelogram 2. The parallelogram tilts, and the length of its sides remains the same.

For nonsquare matrices, it can be shown that there are always some vectors (either rows or columns) that are not independent of the other vectors. Therefore, the determinant of such a nonsquare matrix is always zero; and

3. An important square matrix is the variance–covariance matrix. The **variance–covariance matrix** has variances in its diagonal and covariance in its off-diagonal elements. Thus, the variance–covariance matrix V can be defined as

$$V = \begin{bmatrix} \sigma_x^2 & cov_{xy} \\ cov_{yx} & \sigma_y^2 \end{bmatrix}$$

Because MANOVA may be expressed as the ratio of two multivariate measures of variance, MANOVA may be expressed as the ratio of the determinants of two variance–covariance matrices. That is, the sum-of-squares between-groups, the sum-of-squares-within, and the total sum-of-squares measures used in an ANOVA are replaced by variance–covariance

Parallelogram 1   Parallelogram 2

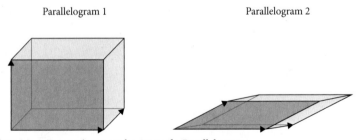

Figure 2.3 A Determinant as the Area of a Parallelogram.

matrices: one matrix for between-groups sum-of-squares (the hypothesis or effect matrix, **H**), one matrix for within-groups sum-of-squares (the error or residual matrix, **E**), and one for total sum-of-squares (the total matrix, **T**). The values in the main diagonal of these matrices are the univariate sums-of-squares for each variable. The other elements are the sums-of-cross products between any two of the variables.

H_0 for a one-way MANOVA is that the population effect of the groups or treatments is zero with respect to all linear combinations of the response variables. This is equivalent to no difference between population centroids (multivariate means). This H_0 can be tested by using a multivariate F-test that is based on the ratio of the multivariate variances of two matrices, such as their determinants. One commonly used multivariate F-test is Wilk's lamda (Λ) (Stevens, 2009; Tabachnick & Fidell, 2007), which is the ratio of the determinants of the sum-of-squares-within-groups matrix (i.e.,$|E|$) and the total sum-of-squares matrix (i.e.,$|T|$) or

$$\frac{|E|}{|T|} \tag{2.6}$$

Wilk's Λ is an inverse criterion: the smaller the value of Λ, the more evidence for the relationship of the IVs with the DVs. If there is no association between the two sets of variables, then Λ would approach 1.

Note that other multivariate F-tests rely on alternative measures of multivariate variance, such as the trace and an eigenvalue. The *trace* of an n-by-n square matrix **A** is defined to be the sum of the elements on the main diagonal (the diagonal from the upper left to the lower right) of **A**. An *eigenvalue* provides quantitative information about the variance in a *portion* of a matrix. Specifically, if **A** is a linear transformation represented by a matrix **A** such that $AX = \lambda X$ for some scalar λ, then λ is called the eigenvalue of **A** with corresponding eigenvector **X**.

Pillai–Bartlett trace is the trace of **H/T** (i.e., the variance between groups). Hotelling–Lawley trace is the ratio of the determinants **H** and **E** or $|H|/|E|$. Roy's largest root is the largest eigenvalue of **H/E** (i.e., the eigenvalue of the linear combination that explains most of the variance and covariance between groups).

The sampling distributions of these statistics are usually converted to approximate F-ratio statistics (Tabachnick & Fidell, 2007), and they will generally produce similar results with more than two groups. The

Pillai–Bartlett trace is the most conservative of these four F-tests, but is a viable alternative if there are reasons to suspect that the assumptions of MANOVA are untenable (Stevens, 2009). In terms of power, none of the aforementioned F-tests is always the choice with the greatest statistical power. Wilk's is the most widely used and consequently more likely to be familiar to readers (Warner, 2008).

Assumptions of MANOVA

MANOVA assumes that data being analyzed follow the GLM (Stevens, 2009; Tabachnick & Fidell, 2007). These assumptions are discussed in further detail in chapter 1, and demonstrated in the annotated example below. In particular, the tenability of following assumptions should be assessed prior to conducting a MANOVA:

1. Multivariate normality of the DVs (robust to moderate violations with respect to type 1 error);
2. Absence of Outliers (sensitive to outliers, especially with a small sample size);
3. Homoscedasticity (robust to moderate violation if groups sizes are approximately equal; largest/smallest < 1.5); and
4. Low to moderate correlation among the DVs.

The MANOVA Procedure

MANOVA may be conceptualized as follows:

1. The omnibus null hypothesis is evaluated with a multivariate F-test;
2. Overall model fit is assessed; and
3. Pairs of group means whose difference are statistically significant are identified.

Evaluating the Omnibus Null Hypothesis

To test the omnibus null hypothesis, multivariate F- tests (discussed earlier in this chapter) are used. These multivariate F-tests include Wilk's Λ, Hotelling's trace, Pillai–Bartlett trace, and Roy's largest root. The purpose of this omnibus hypothesis test, then, is to help to prevent inflating the *family-wise* α-*level*. As discussed in chapter 1, the family-wise α-level increases as the number of separate hypothesis tests increases. Consequently, if separate tests are performed for each DV, the probability of obtaining a false significant value would increase in direct proportion to the number of DVs being tested; that is, the power of the test decreases. If the multivariate F is significant, overall model fit may be assessed, and pairs of means whose difference are statistically significant are identified.

Assessing Overall Model Fit

One measure of overall model fit is an *effect size*. In the context of a one-way MANOVA, an effect size is a retrospective measure of the amount of variability in the set of DVs that are explained by the IV(s). One multivariate measure of effect size is Wilks' λ(discussed above), which reflects the ratio of within-group variance to total variance. That is, as discussed above, Wilk's Λ is an inverse criterion: the smaller the value of Λ, the more evidence for a relationship between the IVs and the DVs. If there is no association between the IVs and the DVs, then Λ would approach 1.

Another measure of effect size for a MANOVA model is *partial eta-squared (η^2)*. As calculated by SPSS partial eta-squared is defined as $1 - $ Wilk's Λ. In SPSS, partial eta square measures the contribution of each factor without controlling for the effects of other factors, and, consequently, it is technically not a true partial measure (Pierce, Block, & Agunis, 2004). In SPSS, the values of partial eta-squared of all DVs can sum to greater than 100%.

Exploring Multivariate Group Differences

The H_0 for a one-way MANOVA is that there are no multivariate group differences (i.e., there are no group differences on linear combinations of the DVs). Accordingly, when conducting a MANOVA, researchers frequently will pose specific questions regarding the nature of multivariate group differences. One issue that is continuously debated by researchers, then, concerns how best to follow-up a statistically significant overall F-test when conducting a MANOVA (Howell, 2009). As discussed earlier, as the number of pair-wise statistical comparisons (e.g., t-tests) increases, the overall type I error rate increases. This problem also may occur when a MANOVA is followed-up with multiple comparisons between groups on linear combinations of the DVs.

There are two basic approaches to assessing multivariate group differences: (1) perform the omnibus test first, followed by a study of some comparisons between pairs of means; or (2) proceed directly to group comparisons and answer specific research questions. That is, some authors (c.f. Huberty & Smith, 1982; Wilcox, 1987) question the need to test the overall null hypothesis, and argue that this second approach may be appropriate when there is an a priori ordering of the DVs, which implies that a specific set of hypotheses be tested. Howell (2009) explains that the hypotheses tested by multivariate F-tests and a multiple comparison tests are different. Multivariate F-tests distribute differences among groups across the number of degrees of freedom for groups. This has the effect of diluting the overall F-value if, for example, several group means are equal to each other but different one other mean. In general, follow-up tests, sometimes termed comparisons or contrasts, are *a priori* or *planned,* or *post hoc* or *unplanned.* An a priori comparison is one that the researcher has decided to test prior to an examination of the data. These comparisons are theory-driven and part of a strategy of confirmatory data analysis. A post hoc comparison is one that a researcher decided to test after observing all or part of the data. These comparisons are data-driven and are part of an exploratory data analysis strategy.

The following comparison strategies are discussed here: (1) multiple ANOVAs (Leary & Altmaier, 1980); (2) two-group multivariate comparisons (Stevens, 2009); (3) step-down analysis (SDA) (Roy & Bargmann, 1958); and (4) simultaneous confidence intervals (SCI) (Smithson, 2003). Multiple ANOVAs and SCIs are univariate perspectives. Two group multivariate comparisons and SDA are multivariate perspectives.

It should be noted that *descriptive discriminant analysis (DDA)* also has been suggested (Huberty, 1994). In DDA, IVs are linearly combined to create a composite IV that maximally differentiates the groups. Although mathematically identical to a one-way MANOVA, DDA emphasizes classification and prediction. The first step in any DDA is to derive *discriminant functions* that are linear combinations of the original variables. The first discriminant function is the linear combination of variables that maximizes the ratio of between-groups to within-groups variance. Subsequent discriminant functions are uncorrelated with previous discriminant functions best separate groups using the remaining variation. The number of discriminant functions that can be extracted equals the number of groups minus one.

MANOVA and DDA may be viewed as providing complementary information. In MANOVA, the focus tends to be on which groups differed significantly. In DDA the focus tends to be on which discriminating variables were useful in differentiating groups. That is, DDA tends to focus on a linear composite of the multiple DVs. It may be argued that if the focus is primarily on the linear composite, DDA may be sufficient, and the researcher should consider using it instead of MANOVA. If, however, the researcher is primarily interested in the individual DVs in the analysis, while controlling for the correlations among the DVs, then MANOVA, may be sufficient. In accordance with MANOVA's emphasis on tests of hypotheses about group differences, subsequent discussion will focus on follow-up group comparison strategies.

Multiple ANOVAs with a Priori or Post Hoc Comparisons. A significant multivariate F-test indicates that there are group differences with respect to their means on the DVs. One popular univariate approach to following up a statistically significant multivariate F-test is to compare planned univariate F-tests and unadjusted group means for each of the DVs.

Another popular variation on the aforementioned univariate strategy is to compare unplanned univariate F-tests and adjusted group means for each of the DVs. Post hoc approaches discussed here include (1) Bonferroni, (2) Tukey honestly significant difference (HSD), and (3) Scheffé. These three tests are a popular subset of possible post hoc corrections, and provide a continuum of choices for conservative to moderate tests of simple and complex comparisons (Klockars, Hancock, & McAweeney, 1995; Kromrey & La Rocca, 1995; Seaman, Levin, & Serlin,1991;Toothacker, 1993).

The Bonferroni procedure. The Bonferroni, a commonly used post hoc test, is relatively simple to compute. The per-comparison α for each individual significance test between means is set as the study-wise α level divided by the number of comparisons to be made (Hsu, 1996). For example, in a one-way MANOVA with five groups $(k = 4)$ and one DV, there are $k (k-1)/2 = 6$ possible pair-wise comparisons of group means. If the researcher wants to limit the overall study-wise type I error rate for the set of six comparisons to .05, then the per-comparison α for each individual significance test between means is the study-wise α level divided by the number of comparisons to be made, or $.05/10 = .005$. The Bonferroni procedure is conservative, and some researchers prefer less conservative methods of limiting the risk of type I error.

The Tukey's HSD (Honestly Significant Difference) test. The Tukey HSD uses a sampling distribution called the **Studentized Range** or **Student's q**, which is similar to a t-distribution (Tukey, 1953). Like t, the q-distribution depends on the number of participants within groups, but the shape of this distribution also depends on k, the number of groups. As the number of groups increases, the number of pair-wise comparisons also increases. To protect against inflated risk of type I error, larger differences between group means are required for rejection of the null hypothesis as k increases.

The distribution of the Studentized range statistic is broader and flatter than the t distribution and has thicker tails. Consequently, the critical values of q are larger than the corresponding critical values of t. The Tukey HSD test could be applied by computing a q value for each comparison of means and then checking to see if the obtained q for each comparison exceeded the critical value of q from a table of the Studentized range statistic.

More specifically, q corresponds to the sampling distribution of the largest difference between two means coming from a set of A means (when $A = 2$, the q distribution corresponds to the usual Student's t). In practice, one computes a criterion denoted $q_{observed}$, which evaluates the difference between the means of two groups. This criterion is computed as:

$$q_{observed} = \frac{M_i - M_j}{\sqrt{MS_{error}\left(\dfrac{1}{s}\right)}} \tag{2.7}$$

where M_i and M_j are the group means being compared MS_{error} is mean square within, which is an estimate of the population variance based on the average of all variances within the several samples, and S is the number of observations per group (the groups are assumed to be of equal size).

Once the $q_{observed}$ is computed, it is then compared with a $q_{critical}$ value from a table of critical values (see this book's companion webpage). The value of $q_{critical}$ depends upon the a-level, the degrees of freedom $v = N_iK$, where N is the total number of participants and K is the number of groups, and a parameter R, which is the number of means being tested. For example, in a group of $K = 5$ means ordered from smallest to largest, $M_1 < M_2 < M_3 < M_4 < M_5$; $R = 5$ when comparing M_5 to M_1; $R = 3$ when comparing M_3 to M_1. The Tukey HSD is considered to be a moderately conservative post hoc test, and is preferred for a large number of simple (i.e., pairwise) comparisons, or for making complex comparisons (Hsu, 1996). A *simple contrast* compares pairs of means. A *complex comparison* compares means to linear combinations of means or linear combinations to mean to each other. For example, a researcher wonders if participants in a control group will experience fewer suicidal ideations compared to the average number of suicidal ideations of participants who received one of three alternative intervention (exp_1, exp_2, and exp_3) strategies. The H_0 for this comparison with contrast coefficients of $(1, -1/3, -1/3, -1/3)$ is as follows:

$$\mu_{\text{wait list}} = \frac{\mu_{exp1} + \mu_{exp2} + \mu_{exp3}}{3} \tag{2.8}$$

The Scheffé Test. The Scheffé test uses F tables versus Studentized range tables (Scheffé, 1953). If the overall null hypothesis is rejected, F-values are computed simultaneously for all possible comparison pairs. These F-values are larger than the value of the overall F test. The formula simply modifies the F-critical value by taking into account the number of groups being compared: $(a - 1) F_{\text{crit}}$. The new critical value represents the critical value for the maximum possible family-wise error rate. This test is the most conservative of all post hoc tests discussed here. Compared to Tukey's HSD, Scheffé has less power when making pair-wise comparisons (Hsu, 1996). That is, although the Scheffé test has the advantage of maintaining the study-wise significance level, it does so at the cost of the increased probability of type II errors. Some authors (cf. Brown & Melamed, 1990) argue that he Scheffé test is used most appropriately if there is a need for unplanned comparisons. Toothacker (1993) recommends the Scheffé test for complex comparisons, or when the number of comparisons is large.

One limitation of using multiple ANOVAs to follow-up a significant multivariate F-test result is that univariate and multivariate tests use different information. If multiple ANOVAs with unadjusted or adjusted mean comparisons are performed, a researcher may assume that if the multivariate null hypothesis is rejected, then at least one of these univariate F-tests will be significant. However, the univariate tests measure the relationship between one DV variable and IV, and the multivariate test uses information about the correlations among the DVs. Consequently, there is no necessary relationship between multivariate significance and univariate significance. That is, rejection of multivariate test does not guarantee that there exists at least one significant univariate F-value (Stevens, 2009). When the correlation among DVs is strong, error on any one DV is accounted for by at least one other DV. As a result, the multivariate error term smaller, and the multivariate test has greater power than the univariate test. This greater power of the multivariate test means that it is more likely to identify a difference among the group means on the DVs using a multivariate versus a univariate test if a difference exists.

   Two-Group Multivariate Comparisons. As outlined by Stevens (1972), the two-group method involves using the Hotelling's T-test statistic (Hotelling, 1931) to simultaneously compare all possible pairs of groups on the set of DVs. As an example, if the MANOVA involved one IV with three levels measured on three DVs, the two-group analysis would involve the multivariate comparison of (1) groups one and two, (2) groups one and three, and (3) groups two and three on the three DVs simultaneously. In general, a significant MANOVA will produce at least one significant T^2. Then, researchers often use univariate t-tests, to explore these statistically significant differences between groups and the DVs. Stevens (1972) compared this approach with DDA, and found that the results, in terms of identifying group differences, were similar, though the two group approach did not allow for direct identification of group differences on individual DVs.

   Stepdown Analysis. Introduced by Roy and Bargmann (1958), stepdown analysis (SDA) is a multistep procedure. Prior to beginning a SDA, the researcher places the DVs in descending order of theoretical importance. It is important to note that this ordering should be based on theoretic issues and not on statistics. Logical orderings of DVs arise when study participants have been tested repeatedly over time, when measures involve progressively more complex behaviors, or whenever there exists a systematic progression from one DV to another.

In step one of a SDA, an analysis of variance (ANOVA) is conducted in which group means on the most important response variable are compared. In step two, the test in step one serves as a covariate and the means of the groups on the second most important DV are compared using an ANCOVA (i.e., in step one there is only one DV in the model). This stepping procedure continues for each DV in turn with DVs at a higher level of importance serving as covariates for those of lesser importance. Testing stops if a significant F-value is encountered, and the H_0 is rejected. Conversely, if no F-value is significant, H_0 is accepted. To control the type I error rate, Tabachnick and Fidell, (2007) recommend the use of a Bonferroni procedure (discussed earlier).

Mudholkar and Subbaiah (1980) cited several potential advantages of using the SDA procedure as a follow up to a significant MANOVA result: (1) simplicity; (2) detailed results for specific variables and groups; (3) useful with small samples; and (4) results for large samples that are equivalent to the omnibus likelihood ratio test. Step-down analysis also has limitations. A potential weakness is the need to order the DVs in terms of theoretical importance. Empirical evidence suggests that the quantitative conclusions reached by researchers can differ substantially depending upon the ordering of DVs (Kowlowsky & Caspy, 1991; Mudholkar & Subbaiah, 1988; Stevens, 1972). In fact, Stevens (2009) argues that a SDA procedure may not be appropriate when a clear a priori ordering of the DVs is not possible. Stevens (1996) also explains that each step in an ordering of DVs is not independent of other steps, and the type I error rate across all of steps in an SDA may not be known. Consequently, the order of DVs may affect the results of an SDA.

Mudholkar and Subbaiah (1980) suggest that when an ordering of DVs is not feasible, SDA may still be used as a post hoc procedure if the observed variables first are subjected to a data-reduction analysis, such as creating linear combinations of DVs with factor analysis. These linear combinations then could be used with for a SDA, with linear combinations of DVs accounting for a greater percentage of the total variance of all DVs in a model placed higher in the sequence. Kowlowsky and Caspy (1991) observed that by applying the SDA to multiple sequences of the response variables a researcher may engage in meaningful data exploration and testing of multiple hypotheses.

Research examining the performance of the SDA procedure, includes a Monte Carlo simulation by Subbaiah and Mudholkar (1978). Using simulated data sets from the multivariate normal distribution Subbaiah and Mudholkar (1978) found that the SDA procedure was able to maintain nominal (or expected) type I error rates (0.01 or 0.05) while attaining power values above 0.8 for most studied conditions. Stevens (1972) used both SDA and discriminant function analysis to compare four groups on eight response variables. The two methods yielded similar results in terms of identifying variables on which the groups differed, while the application of univariate ANOVAs produced a markedly different outcome. According to Finch (2007), when both the assumptions of normality and equality of covariance matrices are not met, the observed type I error rate for the first DV tested in a sequence was inflated, while in all other cases it was near the nominal 0.05 level. This result should serve to encourage researchers using the SDA to assess the viability of both the normality and equality of covariance matrix assumptions prior to making use of the technique.

The essential point of the step-down procedure is that the DVs should be arranged in descending order of importance. For example, with two DVs the **compound** (or **multivariate) hypothesis** is then decomposed into two **component** (or **univariate) hypotheses.** The first component hypothesis concerns the marginal univariate distribution of the most important DV; the second component hypothesis concerns the conditional univariate distribution of the next most important DV given the most important DV. Each of the aforementioned component hypotheses is evaluated by using a univariate F-test. The compound hypothesis is accepted if each of the univariate hypotheses is accepted. The component univariate tests are independent, if the compound hypothesis is true (Roy, 1958).

Simultaneous Confidence Intervals (CIs). A CI is a range of values around which a population parameter (e.g., true mean) is likely to lie in the long run (Neyman, 1952). For example, assuming a normal distribution, if samples of the same size are drawn repeatedly from a population and a 95% CI is calculated around each sample's mean (i.e., plus or minus two standard errors from the mean), then 95% of these intervals should contain the population mean. A second interpretation of a CI is as a significance test (Schmidt & Hunter, 1997). From this perspective, if the value of 0 falls within an interval, the estimate is not statistically significant at the level of the CI.

When multiple CIs are calculated, each with confidence level $1 - \alpha$, the **simultaneous** or **overall confidence level** will be $1 - m\alpha$, where m equals the number of CIs being calculated. Therefore, one approach to controlling the simultaneous confidence level is to set the level of each confidence interval equal to $1 - \alpha/m$. For example, if five CIs are calculated, and the researcher wants to maintain the overall CI at the 95% level, then each of the five CIs would be calculated at the 99% CI. This method of forming simultaneous confidence intervals (SCIs) is called the *Bonferroni method*. Bonferroni-type simultaneous confidence intervals based on the Student's t distribution for the contrast differences across all DVs are available in SPSS, SAS, and Stata (http://www.ats.ucla.edu/stat/stata/examples/alsm/alsmstata17.htm) Alternative approaches to calculating SCIs include Scheffé and Tukey, which are based on the F-distribution and Studentized Range or Student's q-distribution, respectively (discussed further below). According to Maxwell and Delaney (2003), no one comparison method is uniformly best. If all pair-wise comparisons are of interest, Tukey's is the most powerful strategy. If only a subset of pair-wise comparisons is required, Bonferroni is preferred. Using the Scheffé method almost always sacrifices power, unless at least one of the comparisons to be tested is complex. If multiple simultaneous CIs are provided by a software package (e.g., Stata and SPSS), the researcher should select the CI with the narrowest width. Note that although there is no general consensus, a recommended strategy to follow up a significant omnibus test is used to guide the analysis in the annotated example describe below.

Model Validation

In the last stage of MANOVA, the model should be validated. If sample size permits, one approach to validation is sample splitting, which involves creating two subsamples of the data and performing a MANOVA on each subsample. Then, the results can be compared. Differences in results between subsamples suggest that these results may not generalize to the population.

Sample Size Requirements

Statistical power is the ability of a statistical test to detect an effect if the effect exists (Cohen, 1988). That is, power is the probability of rejecting H_0 when a particular alternative hypothesis (H_a) is true. Power can be defined only in the context of a specific set of parameters, and, accordingly, reference to an alternative hypothesis, expressed as an effect size (discussed subsequently), is necessary. The power of a test, therefore, is defined as $1 - \text{beta}$ (β), where β (or type II error) is the probability of falsely accepting H_0 when H_a is true. More specifically, power is a function of α and β, where α (or type I error) is the probability of falsely rejecting H_0 when H_a is false. As α increases, β decreases with a corresponding increase in power. Also, power is a function of effect size (i.e., the true alternative hypothesis, such as an expected difference between groups, how small the effect can be and still be of substantive interest, or the expected size of the effect). As effect size increases, power increases. Finally, power is a function of sample size and variance. In general, the variance of the distribution of an effect size decreases as N increases; and, as this variance decreases, power increases.

The logic behind power analysis is compelling: a study should be conducted only if it relies on a sample size that is large enough to provide an adequate and a prespecified probability of finding an effect if an effect exists. If a sample is too small, the study could miss important differences; too large, the study could make unnecessary demands on respondent privacy and time or waste valuable resources. The benefits of power analysis are at least twofold: sample size selection is made on a rational basis rather than using rules-of-thumb, and the researcher must specify the size of the effect that is substantively interesting.

A widely used measure of effect size in MANOVA is Cohen's (1988) f^2. For multiple correlation and regression, f^2 can be expressed as a function of R^2 (the multiple correlation coefficient):

$$f^2 = \frac{R^2}{1-R^2}. \tag{2.10}$$

However, in the case of MANOVA, the relationship between f^2 and R^2 is complex. Insight into this relationship between f^2 and R^2 is provided in Tables 10.2.1, 10.2.2, and 10.2.3 in Cohen (1988). A review of these tables can assist researchers who are planning studies and who are familiar with R^2 make decisions with regard to the minimum or expected value of f^2.

The strategy for determining sample size recommended here and demonstrated in the annotated example below is to use GPower, which is a free power analysis program available at http://www.psycho.unidues seldorf.de/aap/projects/gpower/.

STRENGTHS AND LIMITATIONS OF MANOVA

MANOVA has the following analytical strengths:

1. By measuring several DVs the researcher improves the chance of discovering what changes as a result of different treatments and their interactions;
2. Protection against inflated type I error as a result of multiple tests of (likely) correlated DVs; and
3. Group differences, not evident in separate ANOVAs, may become apparent. Consequently, MANOVA, which considers DVs in combination, may be more powerful than separate ANOVAs.

MANOVA has the following analytical limitations:

1. The assumption of equality of variance–covariance matrices is more difficult to satisfy than its bivariate analog (equality of variance);
2. There may be ambiguity in interpretation of the effects of IVs on any single DV;
3. The situations in which MANOVA is more powerful than ANOVA are quite limited; often MANOVA is considerably less powerful than ANOVA, particularly in finding significant group differences for a particular DV. Even moderately correlated DVs diminish the power of MANOVA; and
4. When there is more than one DV, there may be redundancy in the measurement of outcomes.

ANNOTATED EXAMPLE

A study is conducted to test a model that compares *satisfaction with agency services* and *satisfaction with parenting* by client *race*. *Satisfaction with agency services* and *satisfaction with parenting* are operationalized as scale scores. Race is operationalized as clients' self-reports of their racial group (coded 1 = Caucasian, 2 = Asian American, 3 = African American). Both the *satisfaction with agency services* and *satisfaction with parenting* scales are considered reliable. Both scales have associated Cronbach's alphas greater than .80. Cronbach's alpha is the most common form of internal consistency reliability coefficient. By convention, a lenient cut-off of .60 is common in exploratory research; alpha should be at least .70 or higher to retain an item in an "adequate" scale; and many researchers require a cut-off of .80 for a "good scale." Cronbach's alpha is discussed further in the section on standard measures and scales, along with other coefficients such as Cohen's kappa (cf. Rubin & Babbie, 2010). Cronbach's alpha can be interpreted as the percentage of variance the observed scale would explain in the hypothetical true scale composed of all possible items in the universe. Alternatively, it can be interpreted as the correlation of the observed scale with all possible other scales measuring the same thing and using the same number of items (Nunnally & Bernstein, 1994).

The data analysis strategy used here in this annotated example is consistent with the recommendations of several prominent authors (cf. Tabachnick & Fidell, 2007; Stevens, 2009) to use Roy–Bargmann stepdown analysis and simultaneous confidence intervals to follow-up comparisons to elucidate a significant MANOVA result. Researchers are strongly encouraged to plan data analysis in detail and include a small number of planned follow-up comparisons after a statistically significant overall F-test. According to Stevens (2009), although "large" is difficult to define, the following considerations mitigate against the use of a large number of DVs: (1) small or negligible differences on many DVs may obscure real differences of a few DVs; (2) the power of the multivariate tests generally declines and the number of DVs is increased; and (3) reliability of variables may be a problem in behavioral science research. It is probably wise to combine highly similar response measures, particularly when basic measurements tend individually to be unreliable.

There are at least three advantages to the proposed strategy. First, the strategy encourages the researcher to think about effects of interest, and to relate these to some underlying theory or to beneficial outcomes of the research. Second, the strategy encourages the researcher to think multivariately, and not to ignore response variable intercorrelations. Third, the researcher gets a clearer perspective on the relative importance on terms of their contributions to group differences.

Accordingly, the researcher decides to perform a one-way MANOVA analysis to investigate the relationship between a client's race (IV), and her/his satisfaction with the adoption agency (DV_1), and satisfaction with parenting (DV_2). Note that throughout this annotated example

SPSS, Stata, and SAS commands are numbered in sequence and highlighted in `Courier`. The data set used in this annotated example, entitled *Annotated_Example1_FULL_N=66.sav*, may be downloaded from this book's companion website. These data are in SPSS format, but may be imported into Stata and SAS.

Determining the Minimally Sufficient Sample Size

GPower, which is a free power analysis program available at http://www. psycho.uni-duesseldorf.de/aap/projects/gpower/, is used to determine the minimally sufficient sample size. For $\alpha = .05$, $\beta = .20$, three groups, two DVs, and $f^2 = 0.15$ (moderate), a total sample size of approximately 66 is adequate.

Prescreening for Multivariate Normality

The tenability of the multivariate normality assumption was evaluated by using an SPSS macro developed by DeCarlo (1997). Open the file *Annotated_Example1_FULL_N=66.sav* in SPSS and type the following commands (numbered in sequence and highlighted in `Courier` into an SPSS's syntax window):

1. `include 'C:\Users\pat\Desktop\MANOVA_Ex_1\`
   `normtest.sps'`
2. `normtest vars = satisfaction_`
   `parenting,satisfaction_adopt_agency`
3. `execute.`

The first line of the aforementioned syntax includes (calls) the macro entitled *normtest.sps*, which is located in the directory *c:\spsswin*, and the second line invokes the macro for the variables *satisfaction_parenting,satisfaction_adopt_agency*. This macro is available from **http://www.columbia.edu/~ld208/** and this book's companion website.

1. Select run → all.

The results of DeCarlo's (1997) are summarized in Figure 2.4. For brevity, the focus here is on Small's (1980) and Srivistava's (1984) tests of

```
Tests of multivariate skew:

    Small's test (chisq)
         Q1          df     ┌─────────┐
                            │ p-value │
     1.7716      2.0000     │  .4124  │
                            └─────────┘

    Srivastava's test
  chi (b1p)         df      ┌─────────┐
                            │ p-value │
     .4892      2.0000      │  .7830  │
                            └─────────┘

Tests of multivariate kurtosis:

    A variant of small's test (chisq)
         VQ2         df     ┌─────────┐
                            │ p-value │
     5.3119      2.0000     │  .0702  │
                            └─────────┘

    Srivastava's test
         b2p       N(b2p)   ┌─────────┐
                            │ p-value │
     2.7672      -.5459     │  .5851  │
                            └─────────┘

    Mardia's test
         b2p       N(b2p)   ┌─────────┐
                            │ p-value │
     7.9961      -.0040     │  .9968  │
                            └─────────┘

Omnibus test of multivariate normality:

    (based on Small's test, chisq)
         VQ3         df     ┌─────────┐
                            │ p-value │
     7.0834      4.0000     │  .1315  │
                            └─────────┘
```

Figure 2.4 Tests of Multivariate Skew.

multivariate kurtosis and skew, Mardia's (1970) multivariate kurtosis, and an omnibus test of multivariate normality based on Small's statistics (see Looney, 1995). However, all other tests calculated by the macro were consistent with these results. That is, all tests of multivariate skew, multivariate kurtosis, and an omnibus test of multivariate normality were not statistically significant at $p = .05$. Each of these tests evaluates the H_0 that the current distribution (these data) equals the multivariate normal distribution. It is hoped that this H_0 will fail to be rejected (accepted). These results suggest that the assumption of multivariate normality is tenable. See chapter 1 for additional information about these tests.

Stata Commands to Prescreen for Multivariate Normality

Mvtest performs multivariate tests of univariate, bivariate, and multivariate normality. All multiple-sample tests provided by mvtest assume independent samples.

1. findit mvtest
2. install
3. mvtest normality satisfaction_parenting satisfaction_adopt_agency, stats(all)

SAS Commands to Prescreen for Multivariate Normality

The *%MULTNORM macro* provides tests and plots of multivariate normality. A test of univariate normality is also given for each of the variables. The macro is available from http://support.sas.com/kb/24/983.html. To use the macro, type the following commands in the SAS editor window:

1. %inc "<location of your file containing the MULTNORM macro>";
2. %multnorm(data=cork, var=n e s w, plot=mult, hires=no)

For this second command statement, DATA = SAS data set to be analyzed. If the DATA = option is not supplied, the most recently created SAS data set is used. VAR = the list of variables to used. Individual variable names, separated by blanks, must be specified. PLOT = MULT requests a high- or low-resolution chi-square quantile–quantile (Q-Q) plot of the squared Mahalanobis distances of the observations from the mean vector. PLOT = UNI requests high-resolution univariate histograms of each variable with overlaid normal curves and additional univariate tests of normality. Note that the univariate plots cannot be produced if HIRES = NO. PLOT = BOTH (the default) requests both of the above. PLOT = NONE suppresses all plots and the univariate normality tests. HIRES = YES (the default) requests that high-resolution graphics be used when creating plots. You must set the graphics device (GOPTIONS DEVICE =) and any other graphics-related options before invoking the macro. **HIRES** = **NO** requests that the multivariate plot be drawn with low-resolution. The univariate plots are not available with `HIRES = NO`. For these data, the following commands were used:

1. `%inc "<location of your file containing the MULTNORM...macro>";`
2. `%multnorm(data=file name, var = satisfaction_parenting satisfaction_adopt_ agency, plot = mult, hires=no)`

Prescreening for Absence of Outliers

The tenability of the absence of outliers assumption was evaluated by using SPSS to calculate Cook's distance, D, which provides an overall measure of the impact of an observation on the estimated MANOVA model. To calculate Cook's D in SPSS, proceed as follows:

1. Click Analyze → General Linear Model → Multivariate
2. Move the DVs to the Dependent Variables box
3. Move the IVs to the Fixed Factor(s) box
4. Click the Model button, and be sure that the radio button Full Factorial is selected
5. Click the Save button and select Cook's distance
6. Click OK

As a result of the aforementioned steps, two variables are added to the data file (one for each DV). These variables are named COO_1 and COO_2. Values for these two variables are the values of D for each DV. Observations with larger values of D than the rest of the data are those which have unusual leverage. Some authors suggest as a cut-off for detecting influential cases values of D greater than $4/(n - k - 1)$, where n is the number of cases and k is the number of independents. Others suggest $D > 1$ as the criterion to constitute a strong indication of an outlier problem, with $D > 4/n$ the criterion to indicate a possible problem. Figure 2.5 summarizes Cook's D cutoffs based on the three

Criterion	Calculated Value for These Data
D greater than $4/(n - k - 1)$	$4/(66\text{-}2\text{-}1) = .06$
$D > 1$	1
$D > 4/n$	$4/66 = .06$

Figure 2.5 Output of DeCarlo's Macro.

aforementioned criteria. For these data, no value of D exceeded .06. These results suggest that the assumption of absence of outliers is tenable.

Stata Commands to Prescreen for Absence of Outliers

1. regress Race satisfaction_parenting satisfaction_adopt_agency
2. _predict cooksd

Cook's D values will be added to the data file as COO_1 and COO_2.

SAS Commands to Prescreen for Absence of Outliers

1. proc reg data="file path";
2. model file Race=satisfaction_parenting satisfaction_adopt_agency;
3. output out= resdata cookd=cooksd
4. run;

Cook's D values will be saved in a new file named "resdata."

Prescreening for Homoscedasticity

MANOVA assumes that for each group (each cell in the factor design matrix) the covariance matrix is similar. Box's M tests the null hypothesis that the observed covariance matrices of the DVs are equal across groups. To calculate Box's M in SPSS, proceed as follows:

1. Click Analyze → General Linear Model → Multivariate

2. Move the DVs to the Dependent Variables box
3. Move the IVs to the Fixed Factor(s) box
4. Click the Model button, and be sure that the radio button Full Factorial is selected
5. Click the Options button and select Estimates of effect size and homogeneity tests
6. Click OK

Researchers do not want M to be significant in order to conclude there is insufficient evidence that the covariance matrices differ. Note, however, that the F-test is quite robust even when there are departures from this assumption. Here M is not significant, and consequently, the assumption of homoscedasticity is tenable (see Figure 2.6).

Stata Commands to Prescreen for Homoscedasticity

```
1. mvtest covariances satisfaction_parenting
   satisfaction_adopt_agency, by(Race)
```

SAS Commands to Prescreen for Homoscedasticity

```
1. proc discrim;
2. class Race;
3. var satisfaction_parenting
   satisfaction_adopt_agency;
4. pool=test wcov;
5. run;
```

Box's M	12.046
F	1.906
df1	6
df2	31697.215
Sig.	.076

Figure 2.6 Box's Test of Equality of Covariance Matrices.

Prescreening for Homogeneity of Regressions

Since a Roy–Bargmann stepdown analysis is planned, data are pre-screened for *homogeneity of regressions*. This assumption, also termed *homogeneity of the regression hyperplanes* or *parallelism*, requires that the regression slope between the covariate and DV is the same (homogeneous) for all groups. According to Stevens (2002), a violation of this assumption means that there is a statistically significant covariate-by-IV interaction. Conversely, if the interaction is not statistically significant this assumption is met.

To test this assumption for a MANOVA with one DV, a model that contains one covariate (i.e., the lower importance DV)-by-IV interaction term is tested for each pair of groups. It is hoped that this interaction term does not achieve statistical significance (e.g., $p > 0.05$). If there is more than one DV, then, one covariate-by-IV interaction term is sequentially included in the model for each DV, and each sequential model is tested for each pair of groups. It is hoped that all covariate-by-IV interaction terms are not statistically significant.

To test the *homogeneity of regressions* for a model that contains a client's race (IV), her/his satisfaction with the adoption agency (DV_1), and satisfaction with parenting (DV_2) proceed as follows:

1. MANOVA satisfaction_adopt_agency by race
   (1,3) with satisfaction_parenting
2. /analysis = satisfaction_adopt_agency
3. /design = satisfaction_parenting,race,
   satisfaction_parenting by race.

1. MANOVA satisfaction_adopt_agency by race
   (1,2) with satisfaction_parenting
2. /analysis = satisfaction_adopt_agency
3. /design = satisfaction_parenting,race,
   satisfaction_parenting by race.

1. MANOVA satisfaction_adopt_agency by race
   (2,3) with satisfaction_parenting
2. /analysis = satisfaction_adopt_agency
3. /design = satisfaction_parenting,race,
   satisfaction_parenting by race.

```
Tests of significance for satisfaction_adopt_agency using
UNIQUE sums of squares
Source of Variation      SS    DF    MS     F   Sig of F

WITHIN+RESIDUAL        173.94  60   2.90
SATISFACTION_PARENTI     2.11   1   2.11   .73     .397
NG
RACE                    13.38   2   6.69  2.31     .108
SATISFACTION_PARENTI    13.39   2   6.69  2.31     .108
NG BY RACE

(Model)                 55.59   5  11.12  3.83     .004
(Total)                229.53  65   3.53

R-Squared =         .242
Adjusted R-Squared = .179
```

Figure 2.7 Group 1 versus Group 3.

In each set of SPSS commands separates the grouping variable (race) from the covariates with the key word with (i.e., with satisfaction_parenting).

These three sets of SPSS commands yield the following three outputs (Figures 2.7, 2.8, and 2.9). Note that the focus of each analysis is on the statistical significance of the interaction between race and satisfaction with parenting (highlighted with a rectangle in the output).

```
Tests of significance for satisfaction_adopt_agency using
UNIQUE sums of squares
Source of Variation      SS    DF    MS     F   Sig of F

WITHIN+RESIDUAL         62.20  31   2.01
SATISFACTION_PARENTI     5.08   1   5.08  2.53     .122
NG
RACE                     6.50   1   6.50  3.24     .082
SATISFACTION_PARENTI     7.42   1   7.42  3.70     .064
NG BY RACE

(Model)                 35.97   3  11.99  5.98     .002
(Total)                 98.17  34   2.89

R-Squared =         .366
Adjusted R-Squared = .305
```

Figure 2.8 Group 1 versus Group 2.

```
Tests of significance for satisfaction_adopt_agency using
UNIQUE sums of squares
Source of Variation       SS   DF    MS       F   Sig of F

WITHIN+RESIDUAL         156.63  46   3.41
SATISFACTION_PARENTI      4.41   1   4.41    1.30    .261
NG
RACE                     13.28   1  13.28    3.90    .054
SATISFACTION_PARENTI     13.11   1  13.11    3.85    .056
NG BY RACE

(Model)                  14.09   3   4.70    1.38    .261
(Total)                 170.72  49   3.48

R-Squared    =      .083
Adjusted R-Squared = .023
```

Figure 2.9 Group 2 versus Group 3.

In each analysis the race and satisfaction with parenting interaction term is not statistically significant at $p = 0.05$. Therefore, homogeneity of regression was achieved for all components of the stepdown analysis.

Stata Commands

1. findit xi3
2. install
3. xi3: regress satisfaction_adopt_agency
   g.race* satisfaction_parenting

SAS Commands

1. proc glm data = file name;
2. class race;
3. model satisfaction_adopt_agency =
   race satisfaction_parenting
   race*satisfaction_parenting;
4. run;
5. quit;

Performing the MANOVA

In SPSS, proceed as follows:

1. Click Analyze → General Linear Model → Multivariate
2. Move the DVs to the Dependent Variables box.
3. Move the IVs to the Fixed Factor(s) box
4. Click the Model button, and be sure that the radio button Full Factorial is selected
5. Click OK

The "Multivariate Tests" section of SPSS's output reports simultaneous tests of each factor's effect on the dependent groups. SPSS offers four alternative multivariate significance tests: Pillai's Trace, Wilk's Lamda, Hotelling's Trace, and Roy's Largest Root. The significance of the F tests show if that effect is significant. For these data, all multivariate tests are statistically significant at $<.05$. The partial eta squared values suggest a low to moderate model fit. The focus here is on Wilk's Lambda, with a partial eta squared value of 0.110. Significant differences were found among the three categories of race on the DVs, with Wilk's Lambda = .793, $F(4,124) = 3.816$, $p <.01$ (see Figure 2.10).

A **Roy–Bargmann stepdown analysis** was conducted to further explore the statistical significance of differences on the DVs by race (see Figure 2.8). It must be noted that the homogeneity of regressions assumption is necessary to perform a Roy–Bargmann stepdown analysis. Recall that according to the above analysis, this assumption is tenable

Multivariate Tests[c]

Effect		Value	F	Hypothesis df	Error df	Sig.	Partial Eta Squared
Intercept	Pillai's Trace	.999	48837.040[a]	2.000	62.000	.000	.999
	Wilks' Lambda	.001	48837.040[a]	2.000	62.000	.000	.999
	Hotelling's Trace	1575.388	48837.040[a]	2.000	62.000	.000	.999
	Roy's Largest Root	1575.388	48837.040[a]	2.000	62.000	.000	.999
Race	Pillai's Trace	.211	3.724	4.000	126.000	.007	.106
	Wilks' Lambda	.793	3.816[a]	4.000	124.000	.006	.110
	Hotelling's Trace	.256	3.904	4.000	122.000	.005	.113
	Roy's Largest Root	.233	7.342[b]	2.000	63.000	.001	.189

a. Exact statistic
b. The statistic is an upper bound on F that yields a lower bound on the significance level.
c. Design: Intercept+Race

Figure 2.10 Multivariate Tests.

for the current model. All DVs were judged to be sufficiently reliable to warrant stepdown analysis, since the reported reliability coefficients for and were and respectively. Both the *satisfaction with agency services* and *satisfaction with parenting* scales are considered reliable. Both scales have associated Cronbach's alphas greater than .80. To conduct a Roy–Bargmann stepdown analysis with two DVs, one ANCOVA is performed with the most import DV as the DV, and the next most important DV as a covariate. (Note that, for example, with three DVs, an ANCOVA is performed with the most important DV as the DV, and the next most important DV as a covariate. Then, an ANCOVA is performed with the next most important DV as the DV, and the next two most important DVs as the covariates.)

To conduct a Roy–Bargmann stepdown analysis with two DVs, proceed as follows in SPSS:

1. Click Analyze → General Linear Model → Univariate
2. Move the DV satisfaction_adopt_agency to the Dependent Variables box.
3. Move the DV satisfaction_parenting to the Covariate(s) box
4. Click the Model button, and be sure that the radio button Full Factorial is selected
5. Click OK

Results of the stepdown analysis are summarized in Figure 2.11. No unique contribution to predicting differences among racial categories was

Dependent Variable: satisfaction_adopt_agency

Source	Type III Sum of Squares	df	Mean Square	F	Sig.
Corrected Model	.088[a]	1	.088	.024	.876
Intercept	219.117	1	219.117	61.120	.000
satisfaction_parenting	.088	1	.088	.024	.876
Error	229.442	64	3.585		
Total	169963.000	66			
Corrected Total	229.530	65			

a. R Squared = .000 (Adjusted R Squared = −.015)

Figure 2.11 Tests of Between-Subject Effects.

made by *satisfaction with agency services* (satisfaction_adopt_agency), $F(1, 66) = 0.024$, $p = .876$.

To further explore differences among racial categories on *satisfaction with agency services*, Table 2.1 displays the mean level of satisfaction with agency services **simultaneous confidence intervals** by racial group. African American respondents report the highest level of satisfaction with agency services followed by Asian American, and Caucasian American respondents. Table 2.1 was produced with the following SPSS commands:

1. Click Analyze → General Linear Model → Multivariate
2. Move the DVs to the Dependent Variables box
3. Move the IVs to the Fixed Factor(s) box
4. Click the Model button, and be sure that the radio button Full Factorial is selected
5. Click the Options button, move Race into the Display Means for window
6. Check Compare main effects
7. For Confidence interval adjustment and select HSD(none)
8. Click Continue
9. Click OK

Stata Commands to Perform a MANOVA

1. manova `satisfaction_parenting satisfaction_ adopt_agency = Race`

Table 2.1 Estimates

Dependent Variable	Race	Mean	Std. Error	95% Confidence Interval	
				Lower Bound	Upper Bound
satisfaction_ parenting	1	50.500	.446	49.609	51.391
	2	49.737	.409	48.919	50.555
	3	50.323	.320	49.682	50.963
satisfaction_ adopt_agency	1	49.312	.431	48.451	50.174
	2	51.000	.396	50.209	51.791
	3	51.258	.310	50.639	51.877

2. anova satisfaction_parenting Race
3. regress
4. anova satisfaction_adopt_agency Race
5. regress

Note that step 3 and 5 calculate simultaneous confidence intervals.

SAS Commands to Perform a MANOVA

In SAS the proc glm procedure uses the following general form:

1. Proc glm;
2. class IV;
3. model DVs = IV;
4. means IV / LSD alpha = p-value;
5. manova h=IV;
6. run;

Note that the aforementioned commands are defined as CLASS <names of the variables to be used as ANOVA factors in the model>; and MODEL <dependent variable(s)> = <independent variable(s)>.

For this analysis, the following commands were used:

1. proc glm;
2. class race;
3. model *satisfaction_parenting satisfaction_ adopt_agency = race;*
4. means race / LSD alpha = .05;
5. manova h=race;
6. run;

Note that Step 4 calculates simultaneous confidence intervals.

SAS Commands to Perform a Roy–Bargmann Stepdown Analysis

1. proc glm;
2. class race;
3. model *satisfaction_adopt_agency = race satisfaction_parenting;*

4. `means race / LSD alpha = .05;`
5. `manova h=race satisfaction_parenting`
   `race*satisfaction_parenting;`
6. `run;`

Stata Commands to Perform a Roy–Bargmann Stepdown Analysis

manova satisfaction_adopt_agency = RACE c.satisfaction_parenting

REPORTING THE RESULTS OF A MANOVA

1. Restate in summary form the reason(s) for the analysis, and the basic components of the model(s) tested including the DV and IVs;
2. Describe how the assumptions underlying the model were tested, and, if not tenable, treated. Include a discussion of missing data, outliers, equality of variance–covariance matrices, and correlation among correlation among DVs;
3. Report the results of at least one multivariate F-test;
4. Report a measure of effect size;
5. If the multivariate F-test is significant, describe how further the statistical significance of differences on the DVs by group were further explored; and
6. Summarize the results of model testing in terms of the study's hypotheses.

Results

This study tested a model that compared a client's satisfaction with an adoption agency's services and their satisfaction with being an adoptive parent by the client's race (coded 1 = Caucasian, 2 = Asian American, 3 = African American). Satisfaction with agency services and satisfaction with parenting were operationalized as scale scores. Race was operationalized as clients' self-reports of their racial group.

A one-way multivariate analysis of variance (MANOVA) was performed to determine the effect of being a member of three racial groups (i.e., African, Asian, and Caucasian American) on a client's satisfaction with an adoption agency's services and their satisfaction with being an adoptive parent.

No problems were noted for missing data, since 0% was missing, outliers, or multivariate normality. The tenability of the absence of outliers assumption was evaluated using Cook's distance, D, which provides an overall measure of the impact of an observation on the estimated MANOVA model. Observations with larger values of D than the rest of the data are those which have unusual leverage. For these data, no value of D exceeded .06. These results suggest that the assumption of absence of outliers is tenable. The tenability of the multivariate normality assumption was evaluated by using an SPSS macro developed by DeCarlo (1997). All tests of multivariate skew, multivariate kurtosis, and an omnibus test of multivariate normality were not statistically significant at $p = .05$. These results suggest that the assumption of multivariate normality is tenable. Box's M was used to test the assumption (i.e., H_0) of equality of variance–covariance matrices. Box's M equaled 12.046, $F(6, 31697) = 1.906$, $p = .076$, which means that equality of variance-covariance matrices can be assumed.

Significant differences were found among the three categories of race on the DVs, with Wilk's Lambda = .793, $F(4,124) = 3.816$, $p < .01$. That is, there are statistically significant differences on the DVs based on race. A Roy–Bargmann stepdown analysis and simultaneous confidence intervals, analysis were conducted to further explore these differences.

For these data, at a statistically significant level, there are mean differences between racial groups on level of satisfaction with the adoption agency (see Table 2.2). At a statistically significant level, African American respondents report the highest level of satisfaction with the adoption

Table 2.2 Means, and Standard Deviations for each Dependent Variable by Race

	Caucasian N = 16 Mean(SD)	Asian American N = 19 Mean(SD)	African American N = 31 Mean(SD)
Satisfaction w/ Adoption Agency	49.31(1.078)*	51.00(1.764)*	51.26(1.949)*
Satisfaction w/ Parenting	50.50(1.826)	40.74(1.485)	50.32(1.922)

*Significant at $p < .05$

agency followed by Asian American, and Caucasian American respondents (see Table 2.2). Although these results are statistically significant, they may not be considered practically significant because group means on the satisfaction agency scale ranged from 49.31 to 51.26.

ADDITIONAL EXAMPLES OF MANOVA FROM THE APPLIED RESEARCH LITERATURE

Colarossi, L. G. (2001). Adolescent gender differences in social support: structure, function, and provider type. *Social Work Research, 25*(4), 233–242.

This study used survey data on 364 adolescents to examine gender differences in perceptions of support across three different constructs: (1) structural support related to the number of adults versus friends; (2) the quantity of support provided by mothers, fathers, peers, and teachers; (3) satisfaction with support from friends and family. Results of a **MANOVA** indicated that young women report a greater number of supportive friends and receive more frequent support from their friends than do young men. However, young men are just as satisfied with friend support as are young women. No gender differences were found in the number of adult supporters, but males received more frequent support from fathers. Nonparental adults emerged as important sources of support for both genders. Implications for practice with adolescents are discussed.

Jewell, J. D., & Stark, K. D. (2003). Comparing the family environments of adolescents with conduct disorder or depression. *Journal of Child and Family Studies, 12*(1), 77–89.

This study attempted to differentiate the family environments of youth with Conduct Disorder (CD) compared to youth with a depressive disorder. Participants were 34 adolescents from a residential treatment facility. The K-SADS-P was used to determine the youth's diagnosis, while their family environment was assessed by the Self Report Measure of Family Functioning Child Version. A *MANOVA* was used to compare the two diagnostic groups on seven family environment variables. Results indicated that adolescents with CD described their parents as having a permissive and ambiguous discipline style, while adolescents with a depressive disorder described their relationship with their parents as enmeshed. A discriminant function analysis, using the two family environment variables of enmeshment and laissez-faire family style as predictors, correctly classified 82% of the participants. Implications for treatment of youth with both types of diagnoses and their families are discussed. (Journal abstract.).

Mowbray, C.; Oyserman, D; Bybee, D.; MacFarlane, P. (2002). Parenting of mothers with a serious mental illness: differential effects of diagnosis, clinical history, and other mental health variables. *Social Work Research*, *26*(4), 225–231.

This study examined the effects of mental illness on parenting in a large urban-based sample of women with serious mental illness. Seventy percent of the sample were women from ethnic minority groups, average age mid-30s; all had care responsibility for at least one minor child. Diagnostic Interview Schedule modules were administered; the women were interviewed to obtain information on parenting, clinical history, and current functioning. MANOVA results suggested that diagnosis had a small but significant negative effect on parenting attitudes and behaviors, and there were race-by-diagnosis interactions. However, current symptoms mediated the effects of diagnosis and chronicity on parenting stress, and current symptomatology and community functioning partially mediated the effects of diagnosis on parenting satisfaction.

O'Hare T. (1998). Alcohol expectancies and excessive drinking contexts in young adults. *Social Work Research*, *22*(1), p44–50.

Despite the growth of alcohol expectancy research, few investigations have examined the association between beliefs in the reinforcing effects of alcohol and situations in which young people drink excessively. The current study of 315 youthful drinkers examined the relationship between three alcohol expectancies (increased social assertiveness, tension reduction, and enhanced sexual pleasure), as measured by the Alcohol Expectancy Questionnaire, and three drinking situation subscales (convivial, personal-intimate, and negative coping) from the Drinking Context Scale. *MANOVA* showed significant main effects for the three expectancy measures. Univariate results showed that expectancies of social assertiveness and tension reduction varied directly with all three excessive drinking contexts. The expectancy of enhanced sexual pleasure, however, varied significantly with personal-intimate drinking only. Practice implications with youthful drinkers are discussed.

Thevos, A. K., Thomas, S. E., & Randall, C. L. (2001). Social support in alcohol dependence and social phobia: Treatment comparisons. *Research on Social Work Practice*, *11*, 458–472.

This study investigated whether different alcoholism treatment approaches differentially impact social support scores in individuals with concurrent alcohol dependence and social phobia. Individuals ($N = 397$) were selected retrospectively from a larger pool of participants enrolled in a multisite randomized clinical trial on treatment matching. Three standard treatments were delivered over 12 weeks: Cognitive-Behavioral Therapy (CBT), Twelve Step Facilitation Therapy (TSF), and Motivational Enhancement Therapy (MET). **MANOVA** was used to analyze social support measures to test the effects of treatment group and gender. For men, there was significant improvement on two measures of social support regardless of treatment group. Women who received CBT or TSF had better support outcomes than women who received MET.

Tsan, J. Y. (2007). Personality and gender as predictors of online counseling use. *Journal of Technology in Human Services, 25*(3), 39–56.

Extraversion, neuroticism, and gender as predictors of online counseling help seeking behavior were investigated. A total of 176 college student participants, 30 men and 146 women, were given the NEO-PI R and ATSPPH-S that assessed, respectively, their personalities and attitudes toward seeking professional psychological help through different modes: traditional face-to-face counseling, video-conferencing, e-mail, instant text message, and microphone. Results were analyzed using **MANOVA** and **MANCOVA**. Subjects were grouped by scores on personality variables: low, medium, and high. Findings suggest that attitudes toward different modes of seeking counseling were associated primarily with gender and extraversion but not with neuroticism.

3

Multivariate Analysis of Covariance

In ANOVA, the mean differences between three or more groups on a single DV are evaluated. ANCOVA assesses group differences on a DV after the effects of one or more covariates are statistically removed. By utilizing the relationship between the covariate(s) and the DV, ANCOVA can increase the power of an analysis.

MANCOVA is an extension of ANCOVA to relationships where a linear combination of DVs is adjusted for differences on one or more covariates. The adjusted linear combination of DVs is the combination that would be obtained if all participants had the same scores on the covariates. That is, *MANCOVA* is similar to MANOVA, but allows a researcher to control for the effects of supplementary continuous IVs, termed covariates. In an experimental design, covariates are usually the variables not controlled by the experimenter, but still affect the DVs. Consequently, although not as effective as random assignment, including covariates may reduce both systematic and within-group error by equalizing groups being compared on important characteristics.

For example, a researcher is planning a study to compare levels of self-esteem and depression among Caucasian, African American, and Mexican American juvenile offenders in a substance-abuse treatment program. Before and after completing the treatment, program participants will be administered a self-esteem scale and a depression scale. It may be useful to adjust the DV scores for pretreatment differences in self-esteem and depression. Here the covariates are pretests of the DVs, a classic use of covariance analysis. After adjustment for pretreatment score differences, post test score differences may be more accurately attributed to treatment.

ASSUMPTIONS OF MANCOVA

For MANCOVA, all of the assumptions for MANOVA apply, with the following additions: (1) covariates are measured without error; (2) a linear relationship between the DV and the covariates; and (3) **homogeneity of the regression hyperplanes**. This assumption, also termed **homogeneity of regressions or parallelism**, requires that the regression slopes between the covariate and DV are the same (homogeneous) for all groups. According to Stevens (2002), a violation of this assumption means that there is a statistically significant covariate-by-IV interaction. Conversely, if the interaction is not statistically significant, this assumption is met.

To test this assumption for a MANCOVA with one covariate, then, a model that contains a covariate-by-IV interaction term is tested. It is hoped that this interaction term does not achieve statistical significance (e.g., $p > .05$). If there is more than one covariate, then, for each covariate, a covariate-by-IV interaction term is included in the model. It is hoped that all covariate-by-IV interaction terms are not statistically significant.

MANCOVA AND MANOVA are similar in terms of (1) evaluating the omnibus null hypothesis, (2) assessing overall model fit, (3) exploring multivariate group differences, and (4) model validation.

SAMPLE SIZE REQUIREMENTS

As in MANOVA, the measure of effect size in MANCOVA is f^2 (Cohen, 1988). Resources for estimating sample size for MANCOVA are difficult to identify. One approach is to adapt the aforementioned sample size estimation strategy for MANOVA. That is, use GPower, which is a free power analysis program available at **http://www.psycho.uniduesseldorf.de/aap/projects/gpower/** and adjust the denominator *df*. If *k* is the number of cells (IVs by DVs) in the design and g is the number of covariates, then groups = $k + g$. This strategy for determining sample size is demonstrated in the annotated example below.

STRENGTHS AND LIMITATIONS OF MANCOVA

MANCOVA has the following analytical strengths:

1. May provide protection against inflated type I error as a result of multiple tests of correlated DVs;
2. Group differences, not evident in separate ANCOVAs, may become apparent;
3. Addition of covariates may reduce systematic bias; and
4. Addition of covariates may reduce within group or error variance.

Regarding analytical strengths three and four, the goal is to obtain maximum adjustment of the DV with minimum loss of degrees of freedom for the error term. Each covariate "costs" one degree of freedom for error. Also, when there is substantial correlation between a covariate and the DV, the gain in terms of power often offsets the loss of power due to reduced degrees of freedom. However, with multiple covariates, a point of diminishing returns is quickly reached, particularly if the covariates are correlated with one another.

In studies with relatively small group sizes (< 20), it is particularly imperative to consider the use of no more than three covariates, because for small or medium effect sizes ($f^2 \leq .15$) power will be low for small group size. Huitema (1980) recommends limiting the number of covariates as follows:

$$\frac{C + (J - 1)}{N} < .10. \tag{3.1}$$

This formula can be re-written as follows:

$$C = (.10N) - (J - 1), \tag{3.2}$$

where C is the number of covariates, J is the number of groups, and N is total sample size. Therefore, for a three-group problem with a total of 60 participants, then $C < 6 - 2 = 4$; that is, four or more covariates are used, then estimates of the adjusted means are likely to be unstable.

All covariates should be correlated with the DV, and none should be substantially correlated with each other. If covariates are substantially correlated with each other, they may not add significantly to reduction of error, or they may cause computational difficulties such as multicollinearity. In addition, to avoid confounding of the intervention effect with a change on the covariate, one should only use pretest or other information gathered before the intervention begins as covariates. If a covariate is used that is measured after the intervention beings and that variable was affected by the intervention, then the change on the covariate may be correlated with change of the DV. Consequently, when the covariate adjustment is made, part of the intervention effect is removed.

MANCOVA has the following analytical limitations:

1. The assumption of equality of variance–covariance matrices is more difficult to satisfy than its bivariate analog of equality of variance;
2. There may be ambiguity in interpretation of the effects of IVs on any single DV;
3. Moderately correlated DVs may diminish the power of MANCOVA; and
4. When there is more than one DV, there may be redundancy in the measurement of outcomes.

ANNOTATED EXAMPLE

A study is conducted to test a model that compares *satisfaction with agency services* and *satisfaction with parenting* by client *race, controlling for client self-efficacy. Satisfaction with agency services, satisfaction with parenting,* and *client self-efficacy* were operationalized as scale scores. Race was operationalized as clients' self-reports of their racial group. The three scales have associated Cronbach's alphas greater than .80. See chapter 2 for a discussion of reliability.

The data analysis strategy used here is consistent with the recommendations of several prominent authors (cf. Tabachnick & Fidell, 2007; Stevens, 2009) to use Roy–Bargmann stepdown analysis and simultaneous confidence intervals to follow-up comparisons to elucidate a significant MANOVA result. See chapter 2 for a discussion of this data analysis strategy.

Accordingly, the researcher decides to perform a one-way MANCOVA to investigate the relationship between a client's *race* (IV), and her/his *satisfaction with the adoption agency* (DV$_1$), and *satisfaction with parenting* (DV$_2$), while controlling for *self-efficacy* (i.e., the covariate). Note that throughout this annotated example SPSS, Stata, and SAS commands are numbered in sequence and highlighted in `Courier`. The data set used in this annotated example, entitled *Annotated_Example1_FULL_N=54. sav*, may be downloaded from this book's companion website. These data are in SPSS format, but may be imported into Stata and SAS.

Determining the Minimally Sufficient Sample Size

GPower, which is a free power analysis program available at http://www. psycho.uni-duesseldorf.de/aap/projects/gpower/ is used to determine the minimally sufficient sample size. The following assumptions are made: $\alpha = .05$, $\beta = .20$, $f^2 = 0.15$ (moderate), the number of groups is three, and the number of DVs equals the number of DVs plus the number of covariates, or three. A total sample size of 54 is adequate.

Prescreening for Multivariate Normality

The tenability of the mulitvariate normality assumption was evaluated by using an SPSS macro developed by DeCarlo (1997). To use the macro, open the file *Annotated_Example1_FULL_N=54.sav* in SPSS, and type the following commands (numbered in sequence and highlighted in `Courier`) into a new SPSS syntax window:

```
1. include 'C:\Users\pat\Desktop\MANCOVA_Ex_1\
   normtest.sps'
2. normtest vars = satisfaction_
   parenting, satisfaction_adopt_agency,
   self_efficacy/.
3. execute.
```

The first line of the aforementioned syntax includes (calls) the macro entitled *normtest.sps*, which is located in the directory *C:\Users\pat*

Desktop\MANCOVA_Ex_1, and the second line invokes the macro for the variables *satisfaction_parenting, satisfaction_adopt_agency, self_effi-cacy*. This macro is available from **http://www.columbia.edu/~ld208/** and this book's companion website.

1. Select run→all.

The results of DeCarlo's (1997) are summarized in Figure 3.1. For brevity, the focus here is on Small's (1980) and Srivistava's (1984) tests of multivariate kurtosis and skew, Mardia's (1970) multivariate kurtosis, and an omnibus test of multivariate normality based on Small's statistics (see Looney, 1995). However, all other tests calculated by the macro were consistent with these results. That is, all tests of multivariate skew,

```
Tests of multivariate skew:

   Small's test (chisq)
         O1          df   p-value
      4.8470      3.0000    .1833

   Srivastava's test
  chi (b1p)          df   p-value
      7.5667      3.0000    .0559

Tests of multivariate kurtosis:

   A variant of small's test (chisq)
         VO2         df   p-value
      6.4722      3.0000    .0908

   Srivastava's test
        b2p    N (b2p)    p-value
      3.2197     .5707      .5682

   Mardia's test
        b2p    N (b2p)    p-value
     15.9375     .6289      .5294

Omnibus test of multivariate normality:

   (based on Small's test, chisq)
         VO3         df   p-value
     11.3192      6.0000    .0790
```

Figure 3.1 Tests of Multivariate Skew.

multivariate kurtosis, and an omnibus test of multivariate normality were not statistically significant at $p = .05$. Each of these tests evaluates the H_0 that the current distribution (these data) equals the multivariate normal distribution. It is hoped that this H_0 will fail to be rejected (accepted). These results suggest that the assumption of multivariate normality is tenable. See chapter 1 for additional information about these tests.

Stata Commands to Prescreen for Multivariate Normality

Mvtest performs multivariate tests of univariate, bivariate, and multivariate normality. All multiple-sample tests provided by mvtest assume independent samples.

1. findit mvtest
2. install
3. mvtest normality satisfaction_parenting
   satisfaction_adopt_agency self_efficacy,
   stats(all)

SAS Commands to Prescreen for Multivariate Normality

The *%MULTNORM macro* provides tests and plots of multivariate normality. A test of univariate normality is also given for each of the variables. The macro is available from http://support.sas.com/kb/24/983.html. To use the macro, type the following commands in the SAS editor window:

1. %inc "<location of your file containing the
   MULTNORM macro>";
2. %multnorm(data=cork, var=n e s w,
   plot=mult, hires=no)

For this second command statement, DATA = SAS data set to be analyzed. If the DATA = option is not supplied, the most recently created SAS data set is used. VAR = the list of variables to used. Individual variable names, separated by blanks, must be specified. PLOT = MULT requests a high- or low-resolution chi-square quantile–quantile (Q-Q) plot of the squared Mahalanobis distances of the observations from the mean vector. `PLOT = UNI` requests high-resolution univariate histograms of each variable with overlaid normal curves and additional univariate tests of normality. Note that the univariate plots cannot be produced if `HIRES = NO`. `PLOT = BOTH` (the default) requests both of the above. `PLOT = NONE` suppresses all plots and the univariate normality tests. `HIRES = YES` (the default) requests that high-resolution graphics be used when creating plots. You must set the graphics device (`GOPTIONS DEVICE =`) and any other graphics-related options before invoking the macro. `HIRES = NO` requests that the multivariate plot be drawn with low-resolution. The univariate plots are not available with `HIRES = NO`.

For these data, the following commands were used:

1. `%inc "<location of your file containing the MULTNORM macro>";`
2. `%multnorm(data=file name, var = satisfaction_parenting satisfaction_ adopt_agency self_efficacy, plot = mult, hires=no)`

Prescreening for Absence of Outliers

The tenability of the absence of outliers assumption was evaluated by using SPSS to calculate Cook's distance, D, which provides an overall measure of the impact of an observation on the estimated MANOVA model. To calculate Cook's D in SPSS, proceed as follows:

1. Click Analyze $\rightarrow$ General Linear Model $\rightarrow$ Multivariate
2. Move the DVs to the Dependent Variables box
3. Move the IVs to the Fixed Factor(s) box
4. Move the covariate(s) to the Covariate(s) box
5. Click the Model button, and be sure that the radio button Full Factorial is selected
6. Click the Save button and select Cook's distance
7. Click OK

Criterion	Calculated Value for These Data
D greater than $4/(n - k - 1)$	$4/(54-2-1) = .07$
$D > 1$	1
$D > 4/n$	$4/54 = .07$

Figure 3.2 Three Citeria for Evaluating Cook's D.

As a result of the aforementioned steps, two variables are added to the data file (one for each DV). These variables are named COO_1 and COO_2. Values for these two variables are the values of D for each DV. Observations with larger values of D than the rest of the data are those which have unusual leverage. Some authors suggest as a cut-off for detecting influential cases, values of D greater than $4/(n - k - 1)$, where n is the number of cases and k is the number of independents. Others suggest $D > 1$ as the criterion to constitute a strong indication of an outlier problem, with $D > 4/n$ the criterion to indicate a possible problem. Figure 3.2 summarizes Cook's D cutoffs based on the three aforementioned criteria. For these data, five cases had values greater than .07 (.08–to .11). If they appear atypical, cases with values of D greater than .07 may be deleted. See chapter 1 for additional discussion about the management of outliers. Alternatively, MANCOVA results should be reported with and without the outliers. Note that for these cases, MANCOVA results with and without cases that had values on one or both DVS that exceeded .07 yielded equivalent results. Consequently, only results for all cases are reported below.

Stata Commands to Prescreen for Absence of Outliers

```
1. regress Race satisfaction_parenting
   satisfaction_adopt_agency self_efficacy
2. _predict cooksd
```

Cook's D values will be added to the data file as coo_1 and COO_2.

SAS Commands to Prescreen for Absence of Outliers

```
1. proc reg data="file path";
```

```
2. model file Race=satisfaction_parenting
   satisfaction_adopt_agency self_efficacy;
3. output out= resdata cookd=cooksd
4. run;
```

Cook's D values will be saved in a new file named "resdata."

Prescreening for Homoscedasticity

MANOVA assumes that for each group (each cell in the factor-design matrix) the covariance matrix is similar. Box's M tests the null hypothesis that the observed covariance matrices of the DVs are equal across groups. To calculate Box's M in SPSS, proceed as follows:

1. Click Analyze → General Linear Model → Multivariate
2. Move the DVs to the Dependent Variables box
3. Move the IVs to the Fixed Factor(s) box
4. Move the covariate(s) to the Covariate(s) box
5. Click the Model button, and be sure that the radio button Full Factorial is selected
6. Click the Options button and select Homogeneity tests
7. Click OK

Researchers do not want M to be significant in order to conclude there is insufficient evidence that the covariance matrices differ. Note, however, that the F-test is quite robust even when there are departures from this assumption. Here M is significant, and, consequently, these suggest that these data exhibit low to moderate heterogeneity ($p = 0.45$). MANCOVA is robust against moderate violations of the homoscedasticity assumption (see Figure 3.3).

Stata Commands to Prescreen for Homoscedasticity

```
mvtest covariances satisfaction_parenting
   satisfaction_adopt_agency self_efficacy,
   by(Race)
```

Box's M	13.658
F	2.147
df1	6
df2	64824.923
Sig.	⟶ .045

Figure 3.3 Box's Test of Equality of Covariance Matrices.

SAS Commands to Prescreen for Homoscedasticity

```
1. proc discrim;
2. class Race;
3. var satisfaction_parenting satisfaction_
   adopt_agency self_efficacy;
4. pool=test wcov;
5. run;
```

Prescreening for Homogeneity of Regressions

Because a Roy–Bargmann stepdown analysis is planned, data are pre-screened for *homogeneity of regressions*. This assumption, also termed *homogeneity of the regression hyperplanes or parallelism*, requires that the regression slope between the covariate and DV is the same (homogeneous) for all groups. According to Stevens (2002), a violation of this assumption means that there is a statistically significant covariate-by-IV interaction. Conversely, if the interaction is not statistically significant this assumption is met.

To test this assumption for a MANOVA with one DV, a model that contains one covariate (i.e., the lower importance DV)-by-IV interaction term is tested for each pair of groups. It is hoped that this interaction term does not achieve statistical significance (e.g., $p > .05$). If there is more than one DV, then, one covariate-by-IV interaction term is sequentially included in the model for each DV, and each sequential model is tested for each pair of groups. It is hoped that all covariate-by-IV interaction terms are not statistically significant.

To test the homogeneity of regressions assumption for a model that contains a client's race (IV), her/his satisfaction with the adoption agency (DV_1), and satisfaction with parenting (DV_2), assuming satisfaction with the adoption agency is logically prior to satisfaction with parenting, proceed as follows:

1. File → New → Syntax
2. Paste the following three blocks of commands into this window:

```
MANOVA satisfaction_adopt_agency by race
   (1,3) with satisfaction_parenting
   self_efficacy
/analysis = satisfaction_adopt_agency
/design = satisfaction_parenting,race,
   satisfaction_parenting by race self_
   efficacy by race.
MANOVA satisfaction_adopt_agency by race
   (1,2) with satisfaction_parenting
   self_efficacy
/analysis = satisfaction_adopt_agency
/design = satisfaction_parenting,race,
   satisfaction_parenting by race self_
   efficacy by race.
MANOVA satisfaction_adopt_agency by race
   (2,3) with satisfaction_parenting
   self_efficacy
/analysis = satisfaction_adopt_agency
/design = satisfaction_parenting,race,
   satisfaction_parenting by race self_
   efficacy by race.
```

3. Run → All

In each set of commands, SPSS separates the grouping variable (race) from the covariates with the keyword **with** (i.e., **with satisfaction_ parenting self_efficacy**).

These three sets of SPSS commands yield the following three outputs (Figures 3.4, 3.5, and 3.6). Note that the focus of each analysis is on the statistical significance of the interaction between race and satisfaction

```
Tests of significance for satisfaction_adopt_agency using
UNIQUE sums of squares
```

Source of Variation	SS	DF	MS	F	Sig of F
WITHIN+RESIDUAL	116.09	46	2.52		
SATISFACTION_PARENTI NG	.00	1	.00	.00	.986
RACE	7.35	2	3.68	1.46	.244
SATISFACTION_PARENTI NG BY RACE	18.28	2	9.14	3.62	.055
SELF_EFFICACY BY RAC E	2.71	2	1.35	.54	.588
(Model)	55.91	7	7.99	3.17	.008
(Total)	172.00	53	3.25		

```
R-Squared =           .325
Adjusted R-Squared =  .222
```

Figure 3.4 Group 1 versus Group 3.

```
Tests of significance for satisfaction_adopt_agency using
UNIQUE sums of squares
```

Source of Variation	SS	DF	MS	F	Sig of F
WITHIN+RESIDUAL	59.98	31	1.93		
SATISFACTION_PARENTI NG	3.76	1	3.76	1.95	.173
RACE	.15	1	.15	.08	.785
SATISFACTION_PARENTI NG BY RACE	5.77	1	5.77	2.98	.094
SELF_EFFICACY BY RAC E	2.54	1	2.54	1.31	.261
(Model)	34.99	4	8.75	4.52	.005
(Total)	94.97	35	2.71		

```
R-Squared =           .368
Adjusted R-Squared =  .287
```

Figure 3.5 Group 1 versus Group 2.

```
Tests of significance for satisfaction_adopt_agency using
UNIQUE sums of squares

Source of Variation          SS     DF      MS       F    Sig of F

WITHIN+RESIDUAL            98.3     31    3.17
SATISFACTION_PARENTI        .06      1     .06     .02      .894
NG
RACE                       6.45      1    6.45    2.03      .164
SATISFACTION_PARENTI      19.23      1   19.23    6.06      .052
NG BY RACE
SELF_EFFICACY BY RAC       1.32      1    1.32     .42      .523
E

(Model)                   21.19      4    5.30    1.67      .182
(Total)                  119.56     35    3.42

RISquared =          .177
Adjusted RISquared = .071
```

Figure 3.6 Group 2 versus Group 3.

with parenting and race and self efficacy (highlighted with rectangles in the output). In each analysis the race and satisfaction with parenting is not statistically significant at $p = .05$. Therefore, homogeneity of regression was achieved for all components of the stepdown analysis.

Stata Commands

1. findit xi3
2. install
3. xi3: regress satisfaction_adopt_
   agency g.race* satisfaction_parenting
   g.race*self_efficacy

SAS Commands

1. proc glm data = file name;
2. class race;
3. model satisfaction_adopt_agency = race
   satisfaction_parenting race*satisfaction_
   parenting race*self_efficacy;
4. run;
5. quit;

Performing the MANOVA

In SPSS, proceed as follows:

1. Click Analyze → General Linear Model → Multivariate
2. Move the DVs to the Dependent Variables box
3. Move the IVs to the Fixed Factor(s) box
4. Move the covariate(s) to the Covariate(s) box
5. Click the Model button, and be sure that the radio button Full Factorial is selected
6. Click OK

The "Multivariate Tests" section reports simultaneous tests each factor's effect on the dependent groups. SPSS offers four alternative multivariate significance tests: Pillai's Trace, Wilk's Lamda, Hotelling's Trace, and Roy's Largest Root. The significance of the F tests show if that effect is significant. For these data, all multivariate tests are statistically significant at $<.05$. The partial eta squared values suggest a low to moderate model fit. The focus here is on Wilk's Lambda, with a partial eta squared value of 0.128. Significant differences were found among the three categories of race on the DVs, with Wilk's Lambda $= .760$, $F(4,98) = 3.600$, $p < .01$ (see Figure 3.7).

A *Roy–Bargmann stepdown analysis* was conducted to further explore the statistical significance of differences on the DVs by race (see Figure 3.8). It must be noted that the homogeneity-of-regressions assumption is necessary to perform a Roy–Bargmann stepdown analysis. Recall that according to the above analysis, this assumption is tenable for the current model. All DVs were judged to be sufficiently reliable to warrant stepdown analysis, since the reported reliability coefficients for and were and respectively. Both the *satisfaction with agency services* and *satisfaction with parenting* scales are considered reliable. Both scales have associated Cronbach's alphas greater than .80. To conduct a Roy–Bargmann stepdown analysis with two DVs, one ANCOVA is performed with the most import DV as the DV, and the next most important DV as a covariate. (Note that, for example, with three DVs, an ANCOVA is performed with most important DV as the DV, and the next most important

Multivariate Tests[C]

Effect		Value	F	Hypothesis df	Error df	Sig.	Partial Eta Squared
Intercept	Pillai's Trace	.734	67.568[a]	2.000	49.000	.000	.734
	Wilks' Lambda	.266	67.568[a]	2.000	49.000	.000	.734
	Hotelling's Trace	2.758	67.568[a]	2.000	49.000	.000	.734
	Roy's Largest Root	2.758	67.568[a]	2.000	49.000	.000	.734
self_efficacy	Pillai's Trace	.060	1.570[a]	2.000	49.000	.218	.060
	Wilks' Lambda	.940	1.570[a]	2.000	49.000	.218	.060
	Hotelling's Trace	.064	1.570[a]	2.000	49.000	.218	.060
	Roy's Largest Root	.064	1.570[a]	2.000	49.000	.218	.060
Race	Pillai's Trace	.241	3.424	4.000	100.000	.011	.120
	Wilks' Lambda	.760	3.600[a]	4.000	98.000	.009	→ .128
	Hotelling's Trace	.314	3.768	4.000	96.000	.007	.136
	Roy's Largest Root	.309	7.734[b]	2.000	50.000	.001	.236

a. Exact statistic
b. The statistic is an upper bound on F that yields a lower bound on the significanc level.
c. Design: Intercept+self_efficacy+Race

Figure 3.7 Multivariate Tests.

DV as a covariate. Then, an ANCOVA is performed with the next most important DV as the DV, and the next two most important DVs as the covariates.)

To conduct a Roy–Bargmann stepdown analysis with two DVs, proceed as follows in SPSS:

1. Click Analyze → General Linear Model → Univariate

Dependent Variable:satisfaction_adopt_agency

Source	Type III Sum of Squares	df	Mean Square	F	Sig.
Corrected Model	40.400[a]	4	10.100	3.761	.010
Intercept	140.908	1	140.908	52.466	.000
self_efficacy	6.634	1	6.634	2.470	→ .122
satisfaction_parenting	.074	1	.074	.027	→ .869
Race	37.085	2	18.542	6.904	.002
Error	131.600	49	2.686		
Total	136978.000	54			
Corrected Total	172.000	53			

a. R Squared = .235 (Adjusted R Squared = .172)

Figure 3.8 Tests of Between-Subjects Effects.

2. Move the DV satisfaction_adopt_agency to the Dependent Variable box
3. Move the IV to the Fixed Factor(s) box
4. Move satisfaction_parenting and self_efficacy to the Covariate(s) box
5. Click the Model button, and be sure that the radio button Full Factorial is selected
6. Click OK

Results of the stepdown analysis are summarized in Figure 2.11. No unique contribution to predicting differences among racial categories was made by *self_efficacy* (self efficacy), $F(1,54) = 2.47$. $p = .122$ and *satisfaction with agency services* (satisfaction_adopt_agency), $F(1, 54) = 0.027$, $p = .869$.

To further explore differences among racial categories on *satisfaction with agency services*, Table 3.1 displays the mean level of satisfaction with agency services with **simultaneous confidence intervals** by racial group. African American respondents report the highest level of satisfaction with agency services followed by Asian American, and Caucasian American respondents. Table 3.1 was produced with the following SPSS commands:

1. Click Analyze → General Linear Model → Multivariate
2. Move the DVs to the Dependent Variables box
3. Move the IVs to the Fixed Factor(s) box
4. Move the covariate(s) to the Covariate(s) box
5. Click the Model button, and confirm that the radio button Full Factorial is selected (this is SPSS's default)
6. Click the Options button, move Race into the Display Means for window
7. Check Compare main effects
8. For Confidence interval adjustment, select HSD(none)
9. Click Continue
10. Click OK

Table 3.1 Estimates

| | | | | Estimates | |
| | | | | 95% Confidence Interval | |
Dependent Variable	Race	Mean	Std. Error	Lower Bound	Upper Bound
satisfaction_ parenting	1	50.4963[a]	.433	49.626	51.366
	2	49.8173[a]	.431	48.951	50.683
	3	50.0203[a]	.443	49.130	50.910
satisfaction_ adopt_agency	1	49.1302[a]	.387	48.353	49.906
	2	50.7633[a]	.385	49.990	51.536
	3	51.1082[a]	.396	50.313	51.902

[a] Covariates appearing in the model are evaluated at the following values: self efficacy = 50.24.

Stata Commands to Perform a MANCOVA

```
1. manova satisfaction_parenting satisfaction_
   adopt_agency = Race c. self_efficacy
2. anova satisfaction_parenting Race
3. regress
4. anova satisfaction_adopt_agency Race
5. regress
```

Note that step 3 and 5 calculate simultaneous confidence intervals.

Stata Commands to Perform a Roy–Bargmann Stepdown Analysis

```
manova satisfaction_adopt_agency = RACE
    c.satisfaction_parenting c.self_efficacy
```

SAS Commands to Perform a MANOVA

In SAS the proc glm procedure uses the following general form:

```
1. Proc glm;
2. class IV;
```

3. model *DVs* = *IV*;
4. means *IV* / LSD alpha = p-value;
5. manova h=*IV*;
6. run;

Note that the options command CLASS is followed by the names of the variables to be used as MANOVA IVs (factors) in the model. The options command MODEL follows the format of names of DV(s) = names of IV(s).

For this analysis, the following commands were used:

1. proc glm;
2. class *race*;
3. model *satisfaction_parenting satisfaction_adopt_agency*
   = *race*;
4. means *race* / LSD alpha =.05;
5. manova h=*race*;
6. run;

Note that Step 4 calculates simultaneous confidence intervals.

SAS Commands to Perform a Roy–Bargmann Stepdown Analysis

1. proc glm;
2. class *race*;
3. model *satisfaction_adopt_agency* = *race*
   satisfaction_parenting;
4. means *race* / LSD alpha =.05;
5. manova h=*race satisfaction_parenting*
   race*satisfaction_parenting;
6. run;

REPORTING THE RESULTS OF A MANCOVA

1. Restate in summary form the reason(s) for the analysis, and the basic components of the model(s) tested including the DV and IVs;

2. Describe how the assumptions underlying the model were tested, and if not tenable treated. Include a discussion of missing data, outliers, equality of variance–covariance matrices, and correlation among correlation among DVs;

3. Report the results of at least one multivariate *F*-test;

4. Report a measure of effect size;

5. If the multivariate *F*-test is significant, describe how further the statistical significance of differences on the DVs by group were further explored; and

6. Summarize the results of model testing in terms of the study's hypotheses.

Results

This study tested a model that compared a client's satisfaction with an adoption agency's services and their satisfaction with being an adoptive parent by the client's race (coded 1 = Caucasian, 2 = Asian, 3 = African American), controlling for client self-efficacy. Satisfaction with agency services, satisfaction with parenting, and client self-efficacy were operationalized as scale scores. Race was operationalized as clients' self-reports of their racial group.

A one-way multivariate analysis of covariance (MANCOVA) was performed to determine the effect of being a member of three racial groups on a client's satisfaction with an adoption agency's services and their satisfaction with being an adoptive parent, controlling for client self-efficacy.

No problems were noted for missing data, outliers, or multivariate normality. The tenability of the absence of outliers assumption was evaluated using Cook's distance, *D*, which provides an overall measure of the impact of an observation on the estimated MANCOVA model. Observations with larger values of *D* than the rest of the data are those which have unusual leverage.

For these data, five cases had values greater than .07 (.08–.11). If they appear atypical, cases that contain values of D greater than 0.07 may be deleted. See chapter 1 for additional discussion about the management of outliers. Alternatively, MANCOVA results should be reported with and without the outliers. Note that for these cases, MANCOVA results with and without cases that had values on one or both DVS that exceeded

.07 yielded equivalent results. Consequently, only results for all cases are reported below. These results suggest that the assumption of absence of outliers is tenable.

The tenability of the multivariate normality assumption was evaluated by using an SPSS macro developed by DeCarlo (1997). All tests of multivariate skew, multivariate kurtosis, and an omnibus test of multivariate normality were not statistically significant at $p = .05$. These results suggest that the assumption of multivariate normality is tenable.

Box's M was used to test the assumption (i.e., H_0) of equality of variance–covariance matrices. Box's M equaled 13.658, $F(6, 64825) = 2.147$, $p = .045$, which means that equality of variance–covariance matrices cannot be assumed.

Significant differences were found among the three categories of race on the DVs, with Wilk's Lambda = .760, $F(4,98) = 3.600$, $p < .01$. A Roy–Bargmann stepdown analysis and simultaneous confidence intervals, analysis were conducted to further explore these differences.

For these data, at a statistically significant level, controlling for client self-efficacy, there are mean differences between racial groups on level of satisfaction with the adoption agency (see Table 3.2). That is, at a statistically significant level, controlling for client self-efficacy. African American respondents report the highest level of satisfaction with the adoption agency followed by Asian American and Caucasian American respondents. Although these results are statistically significant, they

Table 3.2 Means and Standard Deviations for each Dependent Variable by Race

	Caucasian N=18 Mean(SE)	95% CI	Asian N=18 Mean(SE)	95% CI	African American N=18 Mean(SE)	95% CI
Satisfaction w/Parenting	50.50 (.43)	49.26– 51.37	49.82 (.43)	48.95– 50.69	50.02 (.44)	49.13– 50.91
Satisfaction w/Adoption Agency	49.13 (.39)*	48.35– 49.91	50.76 (.39)*	49.99– 51.54	51.11 (.40)*	50.31– 51.90

*Significant at $p < .05$

may not be considered practically significant since group means on the satisfaction with adoption agency scale ranged from 49.13 to 51.10.

ADDITIONAL EXAMPLES OF MANOVA FROM THE APPLIED RESEARCH LITERATURE

Jones, R., Yates, W. R., Williams, S., Zhou, M., & Hardman, L. (1999). Outcome for adjustment disorder with depressed mood: Comparison with other mood disorders. *Journal of Affective Disorders*, 55 (1), 55–61.

Retrospective data were used to evaluate the construct validity of the adjustment disorder diagnostic category. The data primarily consisted of SF-36 Health Status Survey responses by a large group of adult psychiatric outpatients before treatment and again six months after beginning treatment. Respondents were divided into five diagnostic groups, and **MANOVA, MANCOVA** and chi square were used to clarify relationships among diagnoses, sociodemographic data and SF-36 scores.

Ashford, J. A. (2006). Child protective service relationships on parental attitudes in the juvenile dependency process. *Research on Social Work Practice*, 16 (6), 582–590.

This pilot study used **MANCOVA** to examine the attitudes of parents in the child dependency process to determine whether their perceptions of fairness, trustworthiness, and satisfaction with the juvenile dependency system differed across types of relationships: relationships with judges or child protective service (CPS) workers. The study relied on a convenience sample of ($N = 40$) parents who were administered structured interviews with Likert-type items after being in relationships with the authorities for 6 months. The findings supported the study's hypotheses about the nature of the relationships in the family drug court process and the relative contributions of relational versus self-interest factors in explaining variations in parental attitudes.

Diana M. DiNitto, Deborah K. Webb and Allen Rubin. (2002). The Effectiveness of an Integrated Treatment Approach for Clients With Dual Diagnoses, *Research on Social Work Practice*, 12 (5), 621–641.

A **MANCOVA** tested the effectiveness of adding a psychoeducationally oriented group therapy intervention, Good Chemistry Groups, to standard inpatient chemical-dependency services for clients dually diagnosed with mental and substance dependence disorders. Ninety-seven clients were randomly assigned to an experimental group ($n = 48$) and a control group ($n = 49$). Outcome variables included drug and alcohol use, participation in self-help support group meetings, incarceration days, psychiatric symptoms, psychiatric inpatient admissions, compliance with prescribed psychotropic medication plans, and composite scores on the Addiction Severity Index. No significant treatment effects were found on any of the outcome variables.

Farooqi, A., Hägglöf, B., Sedin, G., Gothefors, L., & Serenius, F. (2007). Mental health and social competencies of 10- to 12-year-old children born at 23 to 25 weeks of gestation in the 1990s: A Swedish national prospective follow-up study. *Pediatrics, 120*, 118–133.

The study investigated a national cohort of extremely immature children with respect to behavioral and emotional problems and social competencies, from the perspectives of parents, teachers, and children themselves. **MANCOVA** of parent-reported behavioral problems revealed no interactions, but significant main effects emerged for group status (extremely immature versus control), family function, social risk, and presence of a chronic medical condition, with all effect sizes being medium and accounting for 8% to 12% of the variance. **MANCOVA** of teacher-reported behavioral problems showed significant effects for group status and gender but not for the covariates mentioned above. According to the teachers' ratings, extremely immature children were less well adjusted to the school environment than were control subjects. However, a majority of extremely immature children (85%) were functioning in mainstream schools without major adjustment problems. results. Reports from children showed a trend toward increased depression symptoms compared with control subjects.

Painter, K. (2009). Multisystemic therapy as community-based treatment for youth with severe emotional disturbance. *Research on Social Work Practice, 19*(3), 314–324.

Using **MANCOVA**, this study compares multisystemic therapy (MST) to family skills training combined with case management in community mental health for emotionally disturbed youth. Youth who received MST experienced more improved mental health symptoms, less juvenile justice involvement, and improvement across the linear combination of school functioning, family functioning, mental health symptoms, substance abuse, risk of self-harm, and disruptive or aggressive behavior than did youth who received usual services. Both groups experienced significant improvement in youth functioning.

Pomeroy, E. C., Kiam, R., & Abel, E. M. (1999). The effectiveness of a psycho-educational group for HIV-infected/affected incarcerated women. *Research on Social Work Practice, 9* (2), 171–187.

This study evaluated the effectiveness of a psychoeducational group intervention for HIV/AIDS-infected and affected women at a large southeastern county jail facility. A **MANCOVA** yielded significant differences between the experimental and comparison groups. Subsequent analysis of covariance for each dependent variable indicated significant differences between groups as well. Effect sizes ranged from moderate to strong. The psychoeducational group intervention appeared to be effective in alleviating depression, anxiety, and trauma symptoms among women inmates infected and affected by HIV/AIDS.

Rubin, A., Bischofshausen, S., Conroy-Moore, K., Dennis, K., Hastie, M., Melnick, L., Reeves, D., & Smith, T. (2001). The effectiveness of EMDR in a child guidance center. *Research on Social Work Practice, 11* (4), 435–457.

This study evaluated the effectiveness of adding Eye Movement Desensitization and Reprocessing (EMDR) to the routine treatment regimen of child therapists. MANCOVA found no significant differences in Child Behavior Checklist scores between groups. Subanalyses conducted for 33 clients with elevated pretest scores found moderate effect sizes that approached, but fell short of, statistical significance. These findings raise doubts about notions that EMDR produces rapid and dramatic improvements with children whose emotional and behavioral problems are not narrowly connected to a specific trauma and who require improvisational deviations from the standard EMDR protocol. Further research is needed in light of the special difficulties connected to implementing the EMDR protocol with clients like those in this study.

4

Multivariate Multiple Regression

OVERVIEW AND KEY TERMS

Less frequently termed *canonical regression*, *multivariate multiple regression* (**MMR**) is used to model the linear relationship between more than one IV and more than one DV. MMR is *multiple* because there is more than one IV. MMR is *multivariate* because there is more than one DV. MMR was developed by Bartlett (1938) as an extension of Hotelling's (1935, 1936) canonical correlation analysis. Although the term canonical regression did not appear in this early literature, it was used later by Tintner (1950) and Bartlett (1951) in reference to Bartlett (1938). MMR is a logical extension of OLS regression. MMR estimates the same regression coefficients and standard errors that would be obtained by using separate OLS regression equations for each DV. However, MMR also may be used to test an omnibus H_0 and composite hypotheses for a model. The *omnibus null hypothesis* is that all regression coefficients are equal zero across all DVs. Tests of individual IVs or subsets of IVs across DVs are termed *composite hypotheses* (Cramer & Nicewander, 1979). These composite hypotheses may be also viewed as *sequential* or *hierarchical multiple regressions*. As used here the term hierarchical multiple regression (not to be confused with hierarchical linear models) is similar to stepwise regression, but a researcher, not the computer, determines the order of entry of variables, or sets of variables.

This chapter begins with an introduction to building and refining linear regression models. The remaining discussion focuses on the MMR procedure, and is organized as follows: (1) estimating multivariate regression parameters; (2) testing the omnibus hypothesis; (3) assessing overall model fit; (4) testing composite hypotheses; (5) model validation; (6) sample size requirements; (7) strengths and limitations of MMR; (8) annotated example; (9) reporting the results of a MMR analysis; (10) results of the annotated example; and (11) additional examples from the applied research literature. MMR will be demonstrated with Stata. References to resources for users of SPSS and SAS also are provided.

As described in chapter 1, model building concerns strategies for selecting an optimal set of IVs that explain the variance in one or more DVs (Schuster, 1998). There are three basic types of OLS regression models: fixed, random, and mixed effects (Winer, 1971). A *fixed effect* is one that is assumed to be measured without error. It is also assumed that the values of a fixed effect in one study are the same as the values of the fixed variable in another study. A *random effect* is assumed to be a value drawn from a larger population of values. The values of random effects represent a random sample of all possible values of that effect. Consequently, the results obtained with a random effect can be generalized to all other possible values of that random variable. Because they involve inference, random effects models are less powerful. Random effects models are sometimes referred to as *Model II* or *variance component models*. Analyses using both fixed and random effects are called *mixed effects* models (Rencher, 1998). Discussion in this chapter focuses on the testing of fixed effects models.

Model Refinement: Dependent Variables

After identifying theoretically important variables, and before testing hypotheses about a model, a researcher may wish to maximize parsimony by empirically reducing the number of DVs in the model.

Stepdown analysis (SDA). Introduced by Roy and Bargmann (1958), with this procedure, the DVs in a model are examined sequentially. SDA is recommended by Stevens (2009) and Tabachnick and Fidell (2007) as a follow-up procedure for multivariate analysis of variance (MANOVA).

However, SDA may also be appropriately used as a follow-up to an MMR, if, based on a statistically significant multivariate F-test, the omnibus H_0 is rejected. See chapter 2 for a detailed discussion of SDA.

Model Refinement: Independent Variables

Having identified theoretically important variables, and before proceeding to test hypotheses, a researcher may wish to maximize parsimony by empirically reducing the number of IVs in the model.

Stepwise procedures. These strategies involve identifying regression models in stages. Stepwise procedures were developed before personal computers when time on mainframe computers was at a premium, and when researchers were considering the problem of what to do when there may be more predictors than observations (Cohen & Cohen, 1983). Stepwise procedures include forward selection, backwards elimination, and stepwise regression. These approaches add or remove variables one-at-a-time until some stopping rule is satisfied. Forward selection starts with an empty model. Backward elimination starts with all of the predictors in the model. Stepwise regression is similar to forward selection except that variables are removed from the model if they become nonsignificant as other predictors are added.

More specifically, in stage one of a *forward selection* procedure, the IV best correlated with the DV is included in the equation. In the second stage, the IV with the highest partial correlation with the DV, controlling for the first IV, is entered. This process is repeated, while at each stage controlling for previously entered IVs, until the addition of a remaining IV does not increase R^2 by a statistically significant amount, or until all IVs are entered.

Backward elimination begins with all IVs and removes IVs one at a time until the elimination of one makes a significant difference in R^2. Backward elimination has an advantage over forward selection and stepwise regression because it is possible for a set of variables to have considerable predictive capability even though any subset of them does not. Forward selection and stepwise regression will fail to identify these variable sets. This is because in forward selection or stepwise regression if IVs do not predict well individually, they will not be entered into a model. Backward elimination starts with all IVs in the model, and, consequently, joint predictive capability may be measured.

All-Possible-Regressions. This procedure moves beyond stepwise regression and tests all possible subsets of the set of IVs. If there are k potential independent variables, then there are $2^k - 1$ distinct subsets of variables to be tested. For example, if you have 10 potential independent variables, the number of subsets to be tested is $2^{10} - 1 = 1023$, and if you have 20 potential independent variables, the number is 2^{20}, or more than one million. A variation of all possible regressions is ***best-subsets regression*** in which a researcher can limit the search based on the number and type of variables to be modeled. For example, ***Minitab****'s* (www.Minitab. com) best-subsets procedure identifies the best-fitting regression models that can be constructed with the IVs specified by a researcher. Minitab examines all possible subsets of the predictors, beginning with all models containing one predictor, and then all models containing two predictors, and so on. By default, Minitab displays the two best models for each number of predictors. For example, for a regression model that contains three IVs, Minitab can report the best and second best one-IV models, followed by the best and second best two-IV models, followed by the full model containing all three IVs.

The two most commonly used criteria for ranking all-possible-subsets and best subsets models are ***adjusted R^2*** and the ***Mallows C_p*** statistic (Hocking, 1976; Mallows, 1973). The latter statistic is related to adjusted R^2, but includes a heavier "penalty" for increasing the number of IVs. Defined as

$$\text{Mallows } C_p = \frac{\text{RSS}(p)}{s^2} - (n - 2p) \qquad (4.1)$$

where n is the sample size, p is the number of IVs plus the intercept, RSS(p) is the residual sum-of-squares from a model containing p parameters, and s^2 is the mean residual sum-of-squares from the model containing p parameters. Recall from chapter 2, that residual sum-of-squares (also termed SSW, sum-of-squares-within, and sum-of-squares error) is a measure of the variability within respective groups around that group's mean for the same variable (e.g., age).

The values of C_p are typically positive and greater than one, where lower values are better. Models that yield the best (lowest) values of C_p will tend to be similar to those that yield the best (highest) values of adjusted R^2, but the exact ranking may be slightly different. Compared to adjusted R^2, the C_p criterion tends to favor models with fewer parameters, so it is perhaps more robust to model overfitting. Generally, plots of R^2 and C_p versus the number of variables are examined to determine an optimal model.

SPSS does not compute Mallow's C_p directly from its menu system. However, a syntax file that can be used to calculate C_p may be downloaded from http://www.spsstools.net/Syntax/RegressionRepeatedMeasure/ DoAll-SubsetsRegressions.txt. For Stata users, *rsquare* is a module that will calculate Mallow's C_p. The *rsquare* module can be downloaded from within Stata by typing ***findit rsquare***. For SAS users, the SAS command file, *model_selection.sas*, which can be downloaded from **http://www-personal.umich.edu/~kwelch/workshops/regression/sas/model_selection.sas** calculates C_p.

When all-possible-subsets and best subsets models are reported, it is essential to describe how the model was derived. It is impossible to determine from the numerical results whether a set of IVs was specified before data collection or was obtained by using a selection procedure for finding the "best" model. Parameter estimates and ANOVA tables do not change to reflect which variable selection procedure was used. Results are the same as what would have been obtained if that set of IVs had been specified in advance.

An increasing number of applied researchers believe that computer-controlled model-enhancement procedures, such as stepwise regression and all-possible-subsets regression, are most appropriate in exploratory research. For theory testing, a researcher should base selection of variables and their order of entry into a model on theory, not on a computer algorithm. Menard (1995), for example, writes,

> There appears to be general agreement that the use of computer-controlled stepwise procedures to select variables is inappropriate for theory testing because it capitalizes on random variations in the data and produces results that tend to be idiosyncratic and difficult to replicate in any sample other than the sample in which they were originally obtained. (p. 54)

At each stage in the stepwise process, a computer program must fit a multiple regression model to the variables in the model to obtain an *F*-to-remove value, and the program must fit a separate regression model for each of the variables not in the model in order to obtain their *F*-to-enter values. It should be noted that statistical software packages, such as SPSS, do not fit all models from scratch. Instead, the stepwise search process, for example, is performed by a sequence of transformations to the correlation matrix of the variables in the model. That is, variables are only read in once, and the sequence of adding or removing variables and recalculating the *F*-statistics requires an updating operation on the correlation matrix, which is called "sweeping."

The nominal significance level (e.g., .05) used at each step in stepwise regression is subject to inflation, such that by the last step, the true significance level may be much higher, increasing the chances of type I errors. That is, when many IVs are considered, and there is nothing theoretically compelling about any of them before the data are collected, probability theory suggests that at least one IV will achieve statistical significance (Draper, Guttman, & Lapczak, 1979). Therefore, as more tests are performed, the probability that one or more achieve statistical significance because of chance (type I error) increases. This phenomenon (discussed in chapter 2), sometimes termed probability pyramiding or inflated alpha, explains why IVs that are theoretically unimportant sometimes achieve statistical significance. Stepwise regression, therefore, usually results in measures (*F*-test, *t*-tests, R^2, standard error of the estimate, prediction intervals) that are biased toward too much strength in the relationship between the DV and the IVs. It is incorrect to call them statistically significant because the reported values do not accurately reflect of the selection procedure.

Also troublesome is when there are missing data. Stepwise procedures must exclude observations that are missing for any of the potential IVs. Sometimes one or more of the IVs in the final model are no longer statistically significant when the model is fitted to the data set that includes missing observations that had been deleted, even when these values are missing at random.

Perhaps, the fundamental problem with computer-controlled methods is that they often substitute for thinking about a problem. Statistical computing packages available today do our arithmetic for us in a way that was totally unthinkable thirty years ago. When solving regression equations with many variables could take weeks using a desk calculator, researchers were understandably reluctant to embark on the computational tasks without considerable thought (Searle, 1987). Today, however, with the increasing availability of affordable user-friendly software and powerful personal computers, minimal statistical knowledge is needed for generating what can be voluminous and sophisticated arithmetic. However, this minimal knowledge probably provides an inadequate perspective on what the output of statistical software packages, such as SPSS, means, and how to use it.

Hierarchical multiple regression. What is needed is a strategy that can distinguish between IVs that are better predictors than those that appear to be better predictors because of chance alone. One alternative to computer-driven model testing procedures for independent variables, such as stepwise regression and all-possible-regressions, is **hierarchical multiple regression (HMR).** In HMR a researcher, not the software, determines the order of entry of the variables (Bryk, Raudenbush, Seltzer, & Congdon, 1988). F-tests are used to compute the significance of each added variable (or set of variables) to the explanation reflected in R^2. This hierarchical procedure is an alternative to comparing betas for purposes of assessing the importance of the independents. In more complex forms of hierarchical regression, the model can involve a series of intermediate variables that are dependents with respect to some other independents, but are themselves independents with respect to the ultimate dependent variable. Hierarchical multiple regression, then, can involve a series of regressions for each intermediate as well as for the ultimate dependent variable. As discussed above, in MMR, tests of individual IVs or subsets of IVs across DVs are termed *composite hypotheses* (Cramer & Nicewander, 1979). These composite hypotheses may be also viewed as *sequential or hierarchical multiple regressions.*

THE MULTIVARIATE MULTIPLE REGRESSION PROCEDURE

Estimating Multivariate Regression Parameters

This process is analogous to estimating parameters in OLS regression. Specifically, OLS regression is used to predict the variance in a DV based on linear combinations of IVs. OLS regression also can determine the relative predictive importance of the IVs in a regression model. Power terms can be added as IVs to measure curvilinear relationships with the DV. *Cross-product terms* can be added as IVs to measure *interaction effects*. The estimated values of the IVs (and the value of the y-intercept) can be used to construct a prediction equation.

OLS regression seeks to find the line that best predicts the DV from one or more IVs. This line is sometimes referred to as the **line-of-best-fit**. That is, the goal of regression is to minimize the sum of the squares of the vertical distances of the points from the line-of-best-fit. The line that best predicts the DV from an IV is defined in terms of a slope for each IV and a y-intercept. The **slope** quantifies the gradient of the line, and equals the change in the DV for each unit change in an IV. If the slope is positive, the DV increases as an IV increases. If the slope is negative, the DV decreases as an IV increases. The **y-intercept** is the value of the line when all IVs equal zero. It defines the elevation of the line on the y-axis.

The OLS regression equation for a sample takes the form $y = b_1 x_1 + b_2 x_2 + \cdots + b_n x_n + c$. The b's are the regression coefficients or slopes. The c is the y-intercept. The standardized versions of the b coefficients are the **beta weights**. Within the context of regression analysis, standardization refers to the practice of redefining regression equations in terms of standard deviation units. Standardized coefficients, then, are the estimates resulting from an analysis performed on variables that have been standardized so that they have variances equal to one.

Advocates of standardized regression coefficients point out that the coefficients are the same regardless of an independent variable's underlying scale. Changes of scale are trivial in one sense, for they do not affect the underlying reality or the degree of fit of a linear model to data. Critics of standardized regression coefficients, however, argue that standardization may have an impact on the ability to evaluate the relative importance of different explanatory variables or the relative importance of a given variable in two or more different populations. That is, there is no reason why a change of one SD in one predictor should be equivalent to a change of one SD in another predictor. Some variables are easier to change (e.g., amount of time on the Internet); other variables are more difficult to change (e.g., number of cigarettes smoked); still others are impossible to change (e.g., age). See, for example, Gelman (2008) and (Kim, 1981) for detailed discussions of the strengths and weaknesses of standardized regression coefficients.

Standardized regression coefficients can be calculated in two ways, with both yielding equivalent solutions (Bring, 1994). One way is first to standardize all variables,

$$x_i^* = \frac{x_i - \bar{x}_i}{s_i}, \quad i = 1, \ldots, k,$$

$$y^* = \frac{y_i - \bar{y}}{s_y},$$

(4.2)

where $\bar{x}_i$ and $\bar{y}$ are the means of each variable in the sample and S_i and S_y are the standardized variables. A second way to standardize regression coefficients is by multiplying them by the ratio between the standard deviation of the respective IV (S_i) and the standard deviation of the DV (S_y),

$$B_i = \widehat{\beta}_i = (s_i / s_y),$$

(4.3)

where B_i is the standardized regression coefficient. Traditional sources of detailed discussions multiple OLS regression include (Cohen, Cohen, West, & Aiken, 2003; Keith, 2006).

In MMR, for each DV, there are as many regression coefficients as there are IVs, plus one coefficient for the intercept. Therefore, if there are two DVs and three IVs, there are $2 \times (3 + 1) = 8$ regression coefficients. The estimated IVs, like their OLS regression counterparts, are unbiased and have minimum variance. Finn (1974) provides a succinct mathematical explanation of MMR. Briefly, in the fixed effects regression model, each DV in a sample of n observations may be expressed as a linear function of a set of IVs plus a random error, ε. The number of IVs (x) is denoted by q, and the βs are the regression coefficients as follows:

$$y_1 = \beta_0 + \beta_1 x_{11} + \beta_2 x_{12} + \cdots + \beta_q x_{1q} + \varepsilon_1$$
$$y_2 = \beta_0 + \beta_1 x_{21} + \beta_2 x_{22} + \cdots + \beta_q x_{2q} + \varepsilon_2$$
$$\vdots$$

(4.4)

$$y_n = \beta_0 + \beta_1 x_{n1} + \beta_2 x_{n2} + \cdots + \beta_q x_{nq} + \varepsilon_n$$

Using matrix notation, the aforementioned models for the n observations can be written more concisely as follows:

$$
\begin{pmatrix} y_1 \\ y_2 \\ \vdots \\ y_n \end{pmatrix} = \begin{pmatrix} 1 & x_{11} & x_{12} & \cdots x_{1q} \\ 1 & x_{21} & x_{22} & \cdots x_{2q} \\ \vdots & \vdots & \vdots & \vdots \\ 1 & x_{n1} & x_{n2} & \cdots x_{nq} \end{pmatrix} \begin{pmatrix} \beta_0 \\ \beta_1 \\ \vdots \\ \beta_q \end{pmatrix} + \begin{pmatrix} \varepsilon_1 \\ \varepsilon_2 \\ \vdots \\ \varepsilon_n \end{pmatrix}
\tag{4.5}
$$

or as $\mathbf{Y} = \mathbf{XB} + \varepsilon$. In OLS regression, the single variable y_1 has a unique role in the equation. In MMR, all DVs in the vector $\mathbf{Y}$ are treated analogously, and $\mathbf{Y}$ is the most predictable linear combination of the jointly dependent variable. "Most predictable" may be defined in a least squares sense as minimizing the sum of square residuals in proportion to the variance of the linear combination of the DVS, or in the sense of maximizing the likelihood of the model among all linear combinations of the DVs. See chapter 5 for a discussion of maximum likelihood estimation (MLE).

MMR estimates the same coefficients and standard errors as would be obtained by using separate OLS regressions. MMR examines each DV separately in relation to a linear combination of all predictor variables without imposing any structure across the several resulting regression equations. The IVs are not transformed into mutually uncorrelated variates, and coefficient estimation is identical to estimating regression equations separately for each DV (Cramer & Nicewander, 1979). However, as mentioned above, MMR also may be used to test an omnibus null hypothesis and composite hypotheses for a model (discussed below); it is these two capabilities of MMR that distinguish it from OLS regression.

Testing the Omnibus Null Hypothesis

The omnibus null hypothesis is that all regression coefficients equal zero across all DVs. The purpose of the **omnibus hypothesis** test, then, is to help to prevent inflating the **study-wise alpha level**. If separate tests are performed for each DV, the probability of obtaining a false significant value would increase in direct proportion to the number of DVs being tested; that is, the power of the test decreases. To evaluate the omnibus null hypothesis, multivariate F- tests are used, which include **Wilk's lambda**,

Hotelling's trace, *Pillai's trace*, and *Roy's largest root*. See chapter 2 for a detailed discussion of these multivariate F-tests.

Assessing Overall Model Fit

Goodness-of-fit in MMR is determined in a way that is identical to OLS regression. In MMR, the value of R^2 can be computed for each equation separately to study the effectiveness of each relationship in accounting for observed variation. However, MMR does not yield an overall association measure. The R^2 values (one for each DV) may be correlated, and, consequently, may not indicate the unique variance contribution of each equation to the total set of equations. That is, the R^2 values for each DV may not indicate the unique variance explained for that DV by its set of IVs as a proportion of total variance explained for all DVs.

Put more technically, the R_i^2 statistics are conventional multiple correlation coefficients and reflect assessments of predictive ability (Cramer & Nicewander, 1979). Although the mean of all R_i^2 values is equal to the total redundancy (Muller, 1981), the R_i^2 values for individual regression equations are not analogs of the proportions of variance of the multiple DVs explained by individual redundancy predictor variates, $R_{y|pi}^2$, nor are they analogs of specific canonical correlations squared, R_{ci}^2. Canonical correlation analysis (CCA) seeks to explain the relationship between two or more sets of variables. More specifically, in CCA, the objective is to find linear combinations of the variables within each set such that the two linear combinations of variables have maximum correlation. The several R_i^2 values, which may be found on the diagonal of the matrix product $R_{yx}R_{xx}^{-1}R_{xy}$, are not hierarchically ordered, nor are the values of the DVs predicted by the separate regression equations necessarily uncorrelated.

Testing Composite Hypotheses

If, based on a statistically significant multivariate F-test, the omnibus null hypothesis is rejected, MMR may be used to test IVs across different models (e.g., the DVs). Tests of IVs or subsets of IVs across DVs are termed **composite hypotheses**. That is, MMR also estimates the **between-equation covariances**. Anderson (1999; 2002) describes a methodology for calculating asymptotic variances and covariances for all coefficient estimates in standard canonical correlation analysis under the assumption that the data are generated by a structural linear model with normally distributed variables and disturbances. According to Anderson (1999; 2002), if these conditions hold, the coefficients of the first canonical pair correspond to those in a MMR, and the asymptotic distribution of the sample canonical correlations and coefficients of the canonical variates may be used for statistical inference about the coefficients.

Model Validation

In the last stage of MMR, the model should be validated. If sample size permits, one approach to validation is sample splitting, which involves creating two subsamples of the data and performing an MMR analysis on each subsample. Then, the results can be compared. Differences in results between subsamples suggest that these results may not generalize to the population.

Sample Size Requirements

There is little discussion about the specific sample-size requirements of MMR. However, it is generally agreed that MMR is a special case of canonical correlation analysis (CCA). Stevens (1996) provides a thorough discussion of the sample size for CCA. To estimate the canonical loadings, only for the most important canonical function, Stevens recommends a sample size at least 20 times the number of variables in the analysis. To arrive at reliable estimates for two canonical functions, a sample size of at least 40 – 60 times the number of variables in the analysis is recommended.

Another perspective on estimating sample size for CCA is provided by Barcikowski and Stevens (1975). These authors suggest that CCA may detect stronger canonical correlations (e.g., $R > .7$), even with relatively small samples (e.g., $n = 50$). Weaker canonical correlations (e.g., $R = .3$) require larger sample sizes ($n > 200$) to be detected. Researchers should consider combining both perspectives to triangular on a minimally sufficient sample size for CCA. That is, they should consider the number of canonical functions to be interpreted, and the relative strength of the canonical loadings of the variables represented by the functions of interest.

STRENGTHS AND LIMITATIONS OF MMR

MMR has the following analytical strengths:

1. Efficiency in computation; and
2. Familiarity in interpretation.

MMR has the following analytical limitations:

1. Causality cannot be established with regression models alone;
2. Results assume perfect model specification in terms of all IVs and their functional form (e.g., product term);
3. Sensitivity to outliers, heteroscedasticity of residuals, multicollonearity among IVs, and nonlinearity of relationships between the IVs and the DV.

ANNOTATED EXAMPLE

A study is conducted to test a model that predicts post-adoption service utilization and positive adoption. Specifically, the study tests a model that includes (1) factors influencing the utilization of post-adoption services (*parents' perceptions of self-efficacy, relationship satisfaction between parents*, and *attitudes toward adoption*) as IVs and (2) *service utilization* and positive adoption outcomes (*satisfaction with parenting* and *satisfaction with adoption agency*) as DV. All variables were operationalized as scale scores. The researcher decides to perform a MMR analysis to investigate further the relationship between the following (1) IVs: *parents' perceptions of self-efficacy, relationship satisfaction between parents*, and *attitudes toward adoption*, and (2) the following DVs: *service utilization*, and *satisfaction with parenting*. The next section presents Stata commands, which are numbered in sequence and in Courier. These Stata commands use the format DVs = IVs.

1. ```
manova service_util_i satisfaction_
   parenting = self_efficacy relationship_sat
   attitude_adoption,continuous(self_efficacy
   relationship_sat attitude_adoption)
```

Figure 4.1 illustrates the results of the *manova* command. The overall *F* tests the null hypothesis that regression coefficients for all IVs equal zero for all DVs. The multivariate *F* is based on the sum of squares between
```

```
                Number of obs =      300

                W = Wilks' lambda      L = Lawley-Hotelling trace
                P = Pillai's trace     R = Roy's largest root

     Source  | Statistic    df    F(df1,   df2) =    F      Prob>F
     --------+------------------------------------------------------
      Model  | W  0.2372      3     6.0    590.0    103.57 0.0000 e
             | P  0.9788            6.0    592.0     94.57 0.0000 a
             | L  2.3054            6.0    588.0    112.96 0.0000 a
             | R  1.7993            3.0    296.0    177.53 0.0000 u
     --------+------------------------------------------------------
    Residual |              296
     --------+------------------------------------------------------
 self_effi~y | W  0.9155      1     2.0    295.0     13.62 0.0000 e
             | P  0.0845            2.0    295.0     13.62 0.0000 e
             | L  0.0924            2.0    295.0     13.62 0.0000 e
             | R  0.0924            2.0    295.0     13.62 0.0000 e
     --------+------------------------------------------------------
 relations~t | W  0.8343      1     2.0    295.0     29.29 0.0000 e
             | P  0.1657            2.0    295.0     29.29 0.0000 e
             | L  0.1986            2.0    295.0     29.29 0.0000 e
             | R  0.1986            2.0    295.0     29.29 0.0000 e
     --------+------------------------------------------------------
 attitude_~n | W  0.3593      1     2.0    295.0    263.02 0.0000 e
             | P  0.6407            2.0    295.0    263.02 0.0000 e
             | L  1.7832            2.0    295.0    263.02 0.0000 e
             | R  1.7832            2.0    295.0    263.02 0.0000 e
     --------+------------------------------------------------------
    Residual |              296
     --------+------------------------------------------------------
      Total  |              299

         e = exact,   a = approximate,   u = upper bound on F
```

Figure 4.1   Stata MANOVA Output.

and within groups, and on the sum of crossproducts; that is, it considers correlations between the criterion variables (see chapter two for a more detailed discussion).

The F- and p-values for all four tests under the section labeled "Model," Wilk's lambda, Lawley–Hotelling trace, Pillai's trace, and Roy's largest root, are statistically significant ($p < .001$). Because the overall multivariate tests are significant, it is concluded that there are differences among the DVs as a function of one or more IVs.

> 2. mvreg service_util_i satisfaction_
>   parenting = self_efficacy relationship_sat
>   attitude_adoption

Figure 4.2 illustrates the results of the mvreg command, which first provides the multiple R^2 values for each DV (equation), with associated F- and p-values. Second, the output provides unstandardized regression coefficients, standard errors, t-values, p-values, and 95% confidence intervals for each IV in each model. In Stata `mvreg` is the command used for MMR estimates. The output from the `mvreg` command is similar to the output from the `regress` command, except that there are three equations (one for each DV) instead of one. The coefficients (and all of the output) are interpreted in the same way as they are for any OLS regression. To be clear, the "R-sq" in Figure 4.2 corresponds to the R^2 from the `regress` command, not the adjusted R^2. If regressions were performed for each outcome variable, the same coefficients, standard errors, t- and p-values, and confidence intervals as shown above would be obtained.

MMR does not provide an overall association measure. The R_i^2 statistics are conventional multiple correlation coefficients and reflect assessments of predictive ability. The R_i^2 values predicted by the separate regression equations necessarily uncorrelated. That is, the R^2 values for each DV may not indicate the unique variance explained for those DVs by its set of IVs as a proportion of total variance explained for all DVs.

As mentioned, if a separate regression was run for each DV, the same coefficients, standard errors, t- and p-values, and confidence intervals as shown above would be obtained. The use of the *test* command is one of the compelling reasons for conducting a multivariate regression analysis. One advantage of using the *mvreg* command is that tests of coefficients (IVs) across the DVs may be run. Accordingly, the researcher tests the null hypothesis that the coefficients for the IVs *test self_efficacy, relationship_sat*, and *attitude_adoption* equal 0 in the equations for each of the two DVs.

Equation	obs	Parms	RMSE	"R-sq"	F	P
service_ut~n	300	4	.4099478	0.3366	50.06577	0.0000
satisfacti~g	300	4	2.301917	0.6423	177.1453	0.0000

	Coef.	Std. Err.	t	P>|t|	[95% Conf.	Interval]
service_ut~n						
self_effic~y	-.0252346	.0049885	-5.06	0.000	-.035052	-.0154172
relationsh~t	-.0419408	.0054803	-7.65	0.000	-.052726	-.0311556
attitude)_a~n	-.0028436	.0048637	-0.58	0.559	-.0124154	.0067281
_cons	3.996216	.3564754	11.21	0.000	3.294669	4.697764
satisfacti~g						
self_effic~y	-.0402116	.0280111	-1.44	0.152	-.0953378	.0149146
relationsh~t	-.0193035	.0307725	-0.63	0.531	-.0798641	.041257
attitude)_a~n	.6267131	.0273102	22.95	0.000	.5729664	.6804598
_cons	21.56306	2.001662	10.77	0.000	17.62377	25.50235

Figure 4.2 Results of the Stata's *mvreg* Command.

3. `test self_efficacy relationship_sat`
   `attitude_adoption` (Please see Figure 4.3)

```
(1)   [service_utilization]self_efficacy = 0
(2)   [satisfaction_parenting]self_efficacy = 0
(3)   [service_utilization]relationship_sat = 0
(4)   [satisfaction_parenting]relationship_sat = 0
(5)   [service_utilization]attitude_adoption = 0
(6)   [satisfaction_parenting]attitude_adoption = 0

        F(6, 296) = 113.73
        Prob > F =    0.0000
```

Figure 4.3 Results of the Stata's *test* Command.

4. As a follow-up to the `test` command, the researcher calculated standardized regression coefficients (i.e., *Beta*) to facilitate a comparison of the relative contribution of IVs in the two models (one for each DV).

```
regress service_util_i self_efficacy
   relationship_sat attitude_adoption, beta
regress satisfaction_parenting self_efficacy
   relationship_sat attitude_adoption, beta
```

Figures 4.4 and 4.5 illustrate the results of these two analyses. In Figure 4.4 for the DV *service utilization*, the three variables seem to make equal contributions. In Figure 4.5 for the DV satisfaction with parenting, the variable *attitude toward adoption* seems to make the greatest relative contribution (*Beta* = 0.8011).

SPSS Commands to Perform a MMR

```
GLM service_utilization satisfaction_
   parenting BY self_efficacy relationship_sat
   attitude_adoption
```

service_ut~i	Beta
self_effic~y	−.0530787
relationsh~t	−.0470319
attitude_a~n	.0221673
_cons	.

Figure 4.4 DV is *service utilization*: Standardized Regression Coefficients.

satisfacti~g	Beta
self_effic~y	-.0557719
relationsh~t	-.0244611
attitude_a~n	.8011388
_cons	.

Figure 4.5 DV is satisfaction with parenting: Standardized Regression Coefficients.

```
/METHOD=SSTYPE(3)
/INTERCEPT=INCLUDE
/CRITERIA=ALPHA(.05)
/DESIGN= self_efficacy relationship_sat
   attitude_adoption
```

SAS Commands to Perform a MMR

```
proc reg data = "g:\SAS\hsb2";
model read socst = write math science;
mtest / details print;
run;
quit;
model DV1 DV2 = IV1 + IV2 + IV3
```

REPORTING THE RESULTS OF A MMR ANALYSIS

1. Restate in summary form the reason(s) for the analysis, and the basic components of the model(s) tested including the DV and IVs;
2. Report how the assumptions underlying the model were tested. Describe missing data, and whether it is plausible to assume that they are missing at random. Describe and discuss how (a) outliers were identified and treated; (b) multicollinearity was identified and treated; (c) linearity was identified and treated; and (d) heteroscedasticity was identified and treated;
3. Report the adjusted R^2 for the whole model. Recall that with a large sample, IVs in a model may be statistically significant, but only explain a low proportion of the variability of the DV;
4. Report the regression coefficients, standard errors, and confidence intervals for the IVs. Although PASW will give a negative t-value if the corresponding regression coefficient is negative, you should

drop the negative sign when reporting the results. Degrees of freedom (*df*) for both *F* and *t* values must be given. Usually *df*s are written as subscripts. For *F* the numerator degrees of freedom are given first. *Df*s may also be place in parentheses, or reported explicitly, e.g., $F(3,12) = 4.32$ or $F = 4.32$, $df = 3, 12$. Significance levels can either be reported exactly (e.g., $p = .032$), or in terms of conventional levels (e.g. $p < 0.05$). There are arguments in favor of either approach, but, at a minimum, presentation should be reported in a consistent way. Beware of highly significant *F* or *t* values, whose significance levels will be reported by PASW as, for example, 0.0000. It is an act of statistical illiteracy to write $p = 0.0000$; significance levels can never be zero; distributions such as *t*, *F*, and *chi-square* are asymptotic, and, consequently, the probability observing any value for these distributions, assuming the null hypothesis is true, is never equal to zero. Therefore, significance levels should be presented, for example, as $p < .00005$. *F* and *t* values are conventionally reported to two decimal places, and R^2_{adj} values to the nearest percentage point (sometimes to one additional decimal place). For coefficients, you should be guided by the sample size: for samples in the range of 100 to 1000, three significant figures usually are sufficient; and

5. Summarize the results of model testing in terms of the study's hypotheses.

RESULTS OF THE ANNOTATED EXAMPLE

A multivariate multiple regression analysis was conducted to test a model that predicts post-adoption service utilization and positive adoption outcomes. Specifically, the model includes (1) factors influencing the utilization of post-adoption services (*parents' perceptions of self-efficacy, relationship satisfaction between parents,* and *attitudes toward adoption*) as independent variables and (2) *service utilization* and positive adoption outcomes (*satisfaction with parenting* and *satisfaction with adoption agency*) as dependent variables.

Wilk's lambda, which was used to test the omnibus hypothesis that all beta coefficients across all DVs equal zero, was statistically significant: $F(6, 590) = 103.57$, $p < .001$. Consequently, it was concluded that that one or more independent variables are statistically significant predictors of one or more dependent variables. To further explore the relationships between the independent and each dependent variable, each dependent variable was regressed on all three independent variables. For the model with *service utilization* as the dependent variable, $R^2 = 0.3366$, $F = 50.0677$, $p < .01$. For the model with *satisfaction with parenting* as the dependent variable, $R^2 = 0.6423$, $F = 177.1453$, $p < .01$. These results suggest that the three independent variables are better predictors (i.e., explain more variance) of *satisfaction with parenting* than of *service utilization*.

To facilitate a comparison of the relative contribution of the IVs in the two models, standardized regression coefficients were calculated. For the DV *service utilization*, the three variables seem to make equal contributions (see Figure 4.4). For the DV *satisfaction with parenting*, the variable *attitude toward adoption* seems to make the greatest relative contribution (*Beta* = 0.8011) (see Figure 4.5).

ADDITIONAL EXAMPLES FROM THE APPLIED RESEARCH LITERATURE

Barrera, M., & Garrison-Jones, C. (1992). Family and peer social support as specific correlates of adolescent depressive symptoms. *Journal of Abnormal Child Psychology*, *20*(1), 1–16.

   Family and peer support were distinquished to determine if these sources of support were differentially related to depression symptoms as measured by the Child Assessment Schedule (K. Hodges et al; see record 1982-29404-001). Step-down **multivariate multiple-regression** analyses showed that depression symptoms were uniquely predicted by social relationship variables after accounting for the effects of anxiety and conduct disorder symptoms. Results are consistent with the assertion that social supports significantly contribute to the experience of depressive symptoms by adolescents.

Cargill, B. R., Emmons, K. M., Kahler, C.W., & Brown, R. A. (2001). Relationship among alcohol use, depression, smoking behavior, and motivation to quit smoking with hospitalized smokers. *Psychology of Addictive Behaviors*, *15*(3), 272–275.

   Relationships among depression, alcohol use, and motivation to quit smoking were examined in a sample of hospitalized smokers. **Multivariate multiple regression** and logistic regression analyses indicated that participants with depressed mood were more likely to have a history of problematic drinking.

Participants with depressed mood and a history of problematic drinking were more likely to be nicotine dependent and anticipated greater difficulty refraining from smoking while hospitalized. Overall, depression and alcohol use had stronger associations with smoking-related variables than with smoking cessation motivation variables.

Flores, L. Y., Navarro, R. L., & DeWitz, S. J. (2008). Mexican American high school students' postsecondary educational goals: Applying social cognitive career theory. *Journal of Career Assessment, 16*(4), 489–501.

A ***multivariate multiple regression*** analysis predicting the educational goal aspirations and expectations of Mexican American high school students was examined based on Lent, Brown, and Hackett's Social Cognitive Career Theory and prior research findings with Mexican American samples. No gender or generational status differences were found in educational aspirations or expectations; however, participants reported higher educational aspirations than educational expectations. In addition, results of a multivariate multiple regression analysis suggested that Anglo-oriented acculturation was significantly positively related to educational goal expectations and educational goal aspirations. Mexican-oriented acculturation, college self-efficacy, and college outcome expectations were not significantly related to Mexican American students' educational goals aspirations or expectations. Results are discussed as they relate to improving the educational achievement among Mexican American youth.

Henderson, M. J., Saules, K. K., & Galen, L. W. (2004). The predictive validity of the University of Rhode Island Change Assessment Questionnaire in a heroin-addicted polysubstance abuse sample. *Psychology of Addictive Behaviors, 18*(2), 106–112.

The purpose of this investigation was to examine the predictive utility of the stages-of-change scales of the University of Rhode Island Change Assessment Questionnaire in a heroin-addicted polysubstance-abusing treatment sample. Participants completed the URICA at the beginning of a 29-week treatment period that required thrice-weekly urine drug screens. ***Multivariate multiple regression*** analysis indicated that after controlling for demographic variables, substance abuse severity, and treatment assignment, the stages of change scales added significant variance to the prediction of heroin- and cocaine-free urine samples. The Maintenance scale was positively related to cocaine-free urines and length in treatment. The implications of these findings for treatment and for measuring readiness among individuals using multiple substances while taking maintenance medications are discussed.

Lynch, S. M., & Graham-Bermann, S. A. (2004). Exploring the relationship between positive work experiences and women's sense of self in the context of partner abuse. *Psychology of Women Quarterly, 28*(2), 159–167.

This study examined the relationships among partner abuse, work quality, and women's sense of self. In particular, we explored the potential for women's work to serve as an alternative source of feedback for the self in the context of partner abuse. The sample consisted of working women who reported experiencing a range of partner abuse. Relationships among partner abuse, work quality, and three self-constructs were tested using *multivariate multiple regression*. Work quality was significantly and positively associated with self at work and general self-esteem and approached significance for self at home. There were no significant associations between partner abuse and self at work. Partner abuse was negatively and significantly associated with self at home and approached significance for self-esteem. These varied results support the importance of assessing multiple aspects of the self and the potential of women's work to be a resource in the context of partner abuse.

Miville, M. L., & Constantine, M. G. (2006). Sociocultural predictors of psychological help-seeking attitudes and behavior among Mexican American college students. *Cultural Diversity and Ethnic Minority Psychology, 12*(3), 420–432.

Sociocultural variables of acculturation, enculturation, cultural congruity, and perceived social support were used as predictors of psychological help-seeking attitudes and behaviors among 162 Mexican American college students. *Multivariate multiple regression* analyses indicated that higher cultural congruity, lower perceived social support from family, and higher perceived social support from significant others were significant predictors of positive help-seeking attitudes. In addition, higher acculturation into the dominant society, lower perceived social support from family, and lower perceived social support from friends were significantly predictive of greater help-seeking behavior. Implications for research and practice are discussed.

Miville, M. L., Darlington, P., Whitlock, B., & Mulligan, T. (2005). Integrating identities: The relationships of racial, gender, and ego identities among white college students. *Journal of College Student Development, 46*(2), 157–175.

The authors proposed that racial and gender identities were related to ego identities based on common themes that exist across these different dimensions of identity. A sample of White college students completed the White Racial Identity Attitude Scale, the Womanist Identity Attitude Scale or Men's Identity Attitude Scale, and the Extended Objective Measure of Ego Identity Status. *Multivariate multiple regression* analyses revealed that all ego identity statuses were significantly related to gender and/or racial identity statuses for both women and men. Implications for practice, limitations, and directions for future research are discussed.

Reynolds, A. L., & Constantine, M. G. (2007). Cultural adjustment difficulties and career development of international college students. *Journal of Career Assessment, 15*(3), 338–350.

This study examined the extent to which two dimensions of cultural adjustment difficulties (i.e., acculturative distress and intercultural competence concerns) predicted two specific career development outcomes (i.e., career aspirations and career outcome expectations) in a sample of international college students from Africa, Asia, and Latin America. Although no significant differences among the participants were found by region of origin and gender, *multivariate multiple regression* analyses indicated that higher levels of acculturative distress were predictive of lower levels of career outcome expectations among these international students. Furthermore, findings revealed that greater intercultural competence concerns were predictive of lower career aspirations and lower career outcome expectations. Implications of the findings for career counseling with African, Asian, and Latin American international students are discussed.

Tang, T., & Kim, J. K. (1999). The meaning of money among mental health workers: The endorsement of money ethic as related to organizational citizenship, behavior, job satisfaction, and commitment. *Public Personnel Management, 28*(1), 15–26.

Exploratory and confirmatory factor analyses were conducted to examine the measurement and dimensions of the six-item Money Ethic Scale (MES) using a sample of mental health workers. Results showed that the items of the new MES had very low and negligible cross-loadings and the interfactor correlations were small. Therefore, the three factors (Budget, Evil, and Success) measured fairly independent constructs. In addition, the results of a *multivariate multiple regression* showed that the linear combination of the factors Budget, Evil, and Success was a significant predictor of the linear combination of organizational citizenship behavior, job satisfaction, and organizational commitment.

5

Structural Equation Modeling

OVERVIEW AND KEY TERMS

Structural equation modeling (SEM), also referred to as **causal modeling** and **covariance structure analysis,** is used to evaluate the consistency of substantive theories with empirical data (cf. Bollen, 1989; Guo, Perron, & Gillespie, 2009; Kaplan, 2000; Kline, 2011; Loehlin, 2004; Mulaik, 2009). SEM is appropriate with experimental, nonexperimental, cross-sectional, longitudinal, one-level (non-nested), multilevel (nested), linear, and nonlinear data. SEM is a hybrid model that integrates **path analysis** and **factor analysis.** SEM is related to factor analysis because it may be used to test hypothesized relationships between unmeasured or **latent variables** and **observed or empirical indicators of latent variables.** SEM is related to path analysis because it may be used to test hypothesized relationships between constructs. Thinking of SEM as a combination of factor analysis and path analysis ensures consideration of SEM's two primary components: the measurement model and the structural model.

The **measurement model** describes the relationships between observed variables and the construct or constructs those variables are hypothesized to measure (Bollen, 1989). Confirmatory factor analysis is used in testing the measurement model. A latent variable is defined more accurately to the extent that the measures that define it are strongly related to one another. If, for example, one measure is only weakly correlated with two other measures of the same construct, then that construct will be poorly defined. Ideally, each indicator is a separate measure of the hypothesized latent variable. In SEM, measurement is recognized as error-prone. By explicitly modeling measurement error, SEM seeks to derive unbiased estimates for the relations between latent constructs.

Multi-item scales pose challenges for SEM if all the items are used as indicators of a latent construct (Cattell, 1956). For instance, a model could have too many parameters to estimate relative to the available sample size, resulting in reduced power to detect important parameters. In addition, it might not fit the data sufficiently well because individual items may have less than ideal measurement properties, leading to the rejection of a plausible model.

When describing strategies for incorporating multi-item scales into SEM, it is useful to distinguish among *factors* or *latent variables*, *items* or *observed variables* or *items*, and *groups of items* or *parcels*. Three basic strategies for incorporating lengthy scales into SEM are as follows: (1) including all items individually and (2) combining items (e.g., summed or averaged) into one or more subsets or parcels. The practice of parceling items within scales or subscales has received considerable attention in the structural equation modeling literature (cf. Bandalos, 2002; Little, Cunningham, Shahar, & Widaman, 2002; MacCallum et al., 1999; Nasser & Takahashi, 2003). Item parceling can reduce the dimensionality and number of parameters estimated, resulting in more stable parameter estimates and proper solutions of model fit. When items are severely nonnormal or are coarsely categorized, research suggests item parceling improves the normality and continuity of the indicators and estimates of model fit are enhanced as compared to the original items (Bandalos, 2002).

Item parceling's potential psychometric benefits notwithstanding, this strategy has been controversial. One concern is that parceling results in a loss of information about the relative importance of individual items (Marsh & O'Neill, 1984), because items are implicitly weighted equally in parcels (Bollen &Lennox, 1991). Another concern was that parceling of ordinal scales results in indicators with undefined values, potentially changing the original relations between the indicators and latent variables, for instance, from nonlinear to linear relations (Coanders,

Satorra, & Saris, 1997). Parceling binary or trichotomous items could result in limited range as opposed to the latent trait scale, thereby biasing variance and covariance parameters in SEM (Wright, 1999). Compared with using individual items, parceling could underestimate the relations of the latent variables if the reliability of the scale is low (Shevlin, Miles, & Bunting, 1997).

Studies suggest that single item parcels work best when constructed on unidimensional structures and for items having five response categories (e.g., Bagozzi and Edwards;1998; Kishton & Widaman, 1994; Little et al., 2002; Schau, Stevens, Dauphinee, & Del Vecchio, 1995) parceling is one desirable option. Consequently, to facilitate this introduction to SEM, the scales used in the simulated data annotated example below are assumed to be one-dimensional with high Cronbach's alphas. Consequently, the approach used to incorporate relatively lengthy scales into this example's model is to use the sum of the scale as an indicator of a latent construct.

The **structural model** describes interrelationships among constructs. Equations in the structural portion of the model specify the hypothesized relationships among latent variables.

SEM utilizes two basic types of variables: exogenous and endogenous. **Exogenous variables** are analogous to independent variables. In SEM terminology, other variables regress on exogenous variables. **Endogenous variables** are analogous to dependent variables. A variable that regresses on a variable is always an endogenous variable, even if this same variable is also used as a variable to be regressed on at other points in a model. A variable that is directly observed and measured is called a **manifest or indicator variable.** A variable that is not directly measured is a **latent variable.** The "factors" in a factor analysis are latent variables. Relationships among exogenous variables (latent or manifest) may be described as covariances. **Covariances** are analogous to correlations in that they are defined as nondirectional relationships among independent latent variables. Covariances are unstandardized correlations. That is, covariances are to correlations as unstandardized regression coefficients are to standardized regression coefficients.

Direct effects are relationships among measured and latent variables (Please see Figure 5.1). Direct relationships are **recursive** or **nonrecursive.** A **recursive relationship** is directional and consists of a single path from one variable to another (i.e., $X{\rightarrow}Y$ in Figure 5.1). A **nonrecursive relationship** is mutual or nondirectional.

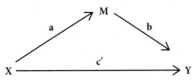

Figure 5.1 A Recursive Relationship.

Mediation in its simplest form represents the addition of a third variable to this $X{\to}Y$ relationship so that it is hypothesized that X causes the mediator, M, and M causes Y, so $X{\to}M{\to}Y$ (Please see Figure 5.1). *Indirect effects* (sometimes referred to as mediated effects) are the relationships between exogenous variables and endogenous variables that are mediated by one or variables. Indirect effects are estimated statistically as the product of direct effects (i.e., ab in Figure 5.1). Mediation may be full or partial. Indirect effects are estimated statistically as the product of direct effects. *Total effects* are the sum of all direct and indirect effects of one variable on another (i.e., $a b + c$ in Figure 5.1).

A simple statistical model consists of exogenous variables whose relationships with a dependent variable are separate. That is, the effects of the exogenous variables are additive, and, consequently, there is additivity, or no interaction. A more complicated model is when the effect of one exogenous variable X_1 on a endogenous variable Y depends on another exogenous variable X_2 (i.e., **moderator**). The relationship between X_1 and X_2 with Y is termed *moderation*. Essentially, *moderator and mediator relationships* are the foundation of structural equation modeling.

When the measurement model and the structural model are considered together, the model is termed the **composite** or *full structural model*. SEM is a largely confirmatory, rather than exploratory, technique. That is, a researcher is more likely to use SEM to determine whether a certain model is valid, rather than use SEM to discover a suitable model. Consequently, because in SEM the researcher is attempting to develop a theoretical model that accounts for all the covariances among the measured items, a nonsignificant difference (accept or fail to reject H_0) between the proposed model and the saturated or perfect model is argued to be suggestive of support for the proposed model. Various *discrepancy functions* can be formed depending on the particular minimization algorithm being used (e.g., maximum likelihood), but the goal remains the same: to derive a test statistic that has a known distribution, and then compare the obtained value of the test statistic against tabled values in order to make a decision about the null hypothesis.

A structural equation model implies a structure of the covariance matrix of the measures (see chapter 2 for a discussion of covariance matrices). Once the model's parameters have been estimated, the resulting model-implied covariance matrix can then be compared to an empirical or data-based covariance matrix. If the two matrices are consistent with one another, then the structural equation model can be considered a plausible explanation for relations between the measures.

From another perspective, assume a set of numbers X related to another set of numbers Y by the equation $Y = 4X$, then the variance of Y *must* be 16 times that of X, so you can test the hypothesis that Y and X are related by the equation $Y = 4X$ *indirectly* by comparing the variances of the Y and X variables. This idea generalizes, in various ways, to several variables interrelated by a group of linear equations. The rules become more complex, the calculations more difficult, but the basic message remains the same: to test whether variables are interrelated through a set of linear relationships by examining the variances and covariances of the variables.

Model fit or goodness-of-fit may be assessed by examining the results of the analysis, in particular the solution (i.e., parameter estimates, standard errors, correlations of parameter estimates, squared multiple correlations, coefficients of determination), the overall fit (i.e., chi-square based and non–chi-squared comparative fit indices), and the detailed assessment of fit (i.e., standardized residuals and modification indices).

The remaining discussion is organized as follows: (1) assumptions, (2) the procedure, (3) sample-size requirements, (4) strengths and limitations, (5) annotated example, (6) reporting results, (7) results, and (8) additional examples of SEM from the applied research literature.

ASSUMPTIONS OF SEM

For SEM, all of the assumptions for MANOVA apply, together with the following extension of the assumption that the model is specified correctly. Because SEM is a confirmatory technique, full model must be defined *a priori*. Within the context of SEM, this assumption, termed **identification**, specifies the requirements of an appropriate model.

As discussed above, thinking of SEM as a combination of factor analysis and path analysis ensures consideration of SEM's two primary components: the measurement model and the structural model. Accordingly, identification concerns whether the parameter of a model (i.e., a set of equations) can be estimated. In SEM, both the structural and the measurement models must be identified. These issues are analogous to the GLM assumptions that a model is correctly specified and measurement error has been minimized, respectively.

Structural Identification

Model identification is a complex topic and a comprehensive mathematical discussion is beyond the scope of this book. However, some insight into identification is essential for researchers to competently perform structural equation modeling. Essentially, the following discussion focuses on the *t*-rule, one of several tests associated with identification. Other tests associated with identification will be briefly described and sources of more comprehensive discussion will be recommended. More extensive discussions of model identification within the context of SEM are provided by Bollen (1989) and Kline (2011).

A statistical model is **structurally identified** if the known information available implies that there is one best value for each parameter in the model whose value is not known. Structural models must be identified for the overall SEM to be identified. That is, a model is identified if the unknown parameters in the model only are functions of identified parameters *and* these functions lead to unique solutions (Bollen, 1989).

Models for which there are an infinite number of possible parameter estimate values are said to be **underidentified**. For example, a theoretical model suggests that X + Y = 10. One possible solution is that X = 5 and Y = 5, another is that X = 2 and Y = 8, but there are many possible solutions for this problem; that is, there is indeterminacy, or the possibility that the data fit more than one implied theoretical model equally well. If a model is underidentified, then it will remain under identified regardless of sample size. Models that are not identified should be respecified.

Models in which there is only one possible solution for each parameter estimate are said to be **just-identified**. Finally, models that have more than one possible solution (but one best or optimal solution) for each parameter estimate are considered **overidentified**. Usually, overidentified

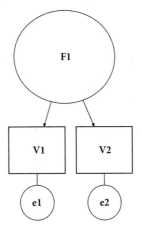

Figure 5.2 An SEM Model.

models are used in SEM because these models allow a researcher to test statistical hypotheses (Loehlin, 2004).

Heuristics are available to help to determine whether a model is structurally identified. One commonly used heuristic is the ***t-Rule***. This rule states that there must be more known pieces of information (i.e., inputs) than unknown pieces of information (i.e., parameters to be estimated) to calculate a unique solution. If this condition is not satisfied, the model is not identified. If this condition is satisfied, the model may be identified. Consider Figure 5.2

This model contains one factor, $F1$, two observed variables, $V1$ and $V2$, and two error variances or residuals, $e1$ and $e2$. This model requires that four parameters be estimated: the factor's variance, the two error variances, and one factor loading. To estimate the number of inputs available to estimate the aforementioned four parameters, use the following formula:

$$[Q(Q+1)] / 2$$

where Q represents the number of measured variables in the model. In this model, there are two observed variables, $V1$ and $V2$, and $[2(2 + 1)]/2 = 3$. It is not possible to estimate four unknown parameters from three inputs. There are three available inputs, but there are four unknown parameters to estimate, and overall, the model has $3 - 4 = -1$ ***degrees of freedom***. This model is underidentified.

Now consider Figure 5.3.

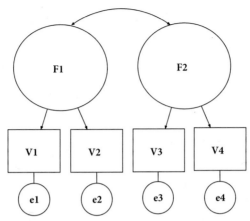

Figure 5.3 An Underidentified SEM Model

This second model has $[4(4 + 1)] / 2 = 10$ available degrees of freedom because there are four observed variables used in the model. This
model has one degree of freedom. This model is structurally identified.
In fact, it is overidentified because there is one positive degree of freedom
present. That is, 10 minus four error variances, two factor loadings, two
factor variances, and one covariance between the factors.

The t-Rule provides necessary, but not sufficient, conditions for
identification. That is, if the t-Rule test is not passed, the model is not
identified. If the test is passed, the model still may not be identified.
Other heuristics to help to determine whether a model is identified
include the Null B Rule, the Recursive Rule, the Order Condition Test,
and the Rank Condition Test. The **Null B Rule** applies to models in
which there is no relationship between the endogenous variables, which
is rare. The Null B Rule provides sufficient, but not necessary conditions for identification. The **Recursive Rule** is sufficient but not necessary for identification. To be recursive, there must not be any feedback
loops among endogenous variables. The **Order Condition Test** suggests that the number of variables excluded from each equation must
be at least $P - 1$, where P is the number of equations. If equations
are related, then the model is underidentified. This is a necessary, but
not a sufficient condition to claim identification. The **Rank Condition
test** is both necessary and sufficient, making it one of the better tests
of identification. Passing this test means the model is appropriate for
further analyses.

SEM software programs such as AMOS perform identification checks as part of the model-fitting process. They usually provide warnings about underidentification conditions. AMOS, for example, displays the following message: The specified model is probably unidentified. In order to achieve identifiability, it will probably be necessary to impose additional constraints. The (probably) unidentified parameters are marked. The method that AMOS uses for determining that a model is unidentified, and for determining how many additional constraints are required to make the model identified, is fallible. However, it is usually right. In order to decide whether a parameter is identified, or whether an entire model is identified, AMOS examines the rank of the matrix of approximate second derivatives, and of some related matrices. There are objections to this approach in principle (Bentler & Weeks, 1980; McDonald, 1982). There are also practical problems in determining the rank of a matrix in borderline cases. Because of these difficulties, identifiability of a model should be judged *a priori*. With complex models, this may be impossible, so a researcher will have to rely on AMOS's numerical determination.

Empirical Identification

Measurement models must also be ***empirically identified*** for the overall SEM to be identified (Kenny & Judd, 1986). A model in which at least one parameter estimate is unstable is ***empirically underidentified***. As discussed above, in SEM, the measurement model describes the relationships between observed variables and the construct or constructs those variables are hypothesized to measure. The measurement model of SEM allows the researcher to evaluate how well his or her observed (measured) variables combine to identify underlying hypothesized constructs. Confirmatory factor analysis is used in testing the measurement model, and the hypothesized factors are referred to as ***latent variables.*** The measures chosen by the researcher define the latent variables in the measurement model. A latent variable is defined more accurately to the extent that the measures that define it are strongly related to one another. Accordingly, in building measurement models, multiple-indicator measurement models (Hunter & Gerbing, 1982) are preferred because they allow the most unambiguous assignment of meaning to the estimated constructs. The reason for this is that with multiple-indicator measurement models, each estimated construct is defined by at least two measures, and each measure is intended as an estimate of only one construct. In contrast, measurement models that contain correlated measurement errors or that have indicators that load on more than one estimated construct do not represent unidimensional construct measurement. As a result, assignment of meaning to such estimated constructs can be problematic.

More specifically, for a measurement model to be empirically identified, latent variables must be scaled. Scaling the latent variable creates one less unknown. Because latent variables are unobserved, they do not have a predefined unit of measurement; consequently, the researcher needs to set the unit of measurement (Brown, 2006). There are two ways to do this. One option is to make it the same as that of one of the indicator variables. The second option is to set the variance equal to 1 for the latent variable. In general, the first option is the more popular. Scaling the latent variable has been compared to converting currency (Brown, 2006). For example, in creating a latent variable for cost of living across the United States, United Kingdom, and France, a researcher has three indicators—one in U.S. dollars, one in British pounds, and the other in Euros. Dollars, pounds, and Euros all have different scales of measurement, but the latent variable can be scaled to any one of these. If scaled to U.S. dollars, the latent variable will be interpretable in terms of dollars. But, the latent variable could also be scaled to either pounds or Euros—whichever will be most interpretable and meaningful for the intended audience.

In addition, for a measurement model to be empirically identified at least one of the following three conditions must hold for each construct in the model:

1. The construct has at least three indicators whose errors are uncorrelated with each other. Note that correlated error terms in measurement models represent the hypothesis that the unique variances of the associated indicators overlap; that is, they measure something in common other than the latent constructs that are represented in the model. Another way to describe an error correlation is as an unanalyzed association, which means that the specific nature of the shared "something" is unknown. Correlated within-factor measurement errors may imply, for example, (a) the presence of another common factor and (b) direct causal relations among indicators (e.g., a response bias set

where one item partly "causes" the response to the next item in a survey);

2. The construct has at least two indicators whose errors are uncorrelated and either (a) both the indicators of the construct correlate with a third indicator of another construct but neither of the two indicators' errors is correlated with the error of that third indicator, or (b) the two indicators' loadings are set equal to each other; and

3. The construct has one indicator and (a) its error variance is fixed to zero or some other *a priori* value (e.g., the quantity one minus the reliability times the indicator's variance), or (b) there is a variable that can serve as an instrumental variable in the structural model and the error in the indicator is not correlated with that instrumental variable. Note that for a variable to be a valid instrument, then, it must be (a) correlated with the dependent variable of a model and (b) only affect the dependent variable through an independent variable.

The remedy for all forms of underidentification is to try to locate the source of the identification problem and determine if the source is empirical underidentification or structural underidentification. For structural underidentification, the only remedy is to respecify the model. Empirical underidentification may be correctable by collecting more data or respecifying the model. (Hayduk 1996), for example, suggests that we can learn a good bit about whether a model is identified by working with it. If the software fails to converge to a solution, or matrices cannot be inverted, the model is probably underidentified. If the model is estimated, but produces absurd results, it may be underidentified. A researcher also may estimate a model by specifying different starting values for MLE parameter estimates. If consistent results are obtained, then the model is probably identified. Note that this will be discussed in more detail below.

THE SEM PROCEDURE

Model Specification

Often, the most difficult part of SEM, model specification involves determining every relationship and parameter in the model that is of interest to the researcher. That is, prior to any data collection or analysis, the researcher describes a model to be confirmed. Available information is used to decide which variables to include in the theoretical model, which implicitly also involves which variables not to include in the model and how these variables are related. A model will be misspecified to the extent that the relationships hypothesized do not capture the observed relationships. The following basic rules are used when drawing a model:

1. Latent variables are represented with circles and measured variables are represented with squares;
2. Lines with an arrow in one direction show a hypothesized direct relationship between the two variables. It should originate at the causal variable and point to the variable that is caused;
3. Absence of a line indicates there is no causal relationship between the variables;
4. Lines with an arrow in both directions should be curved and this demonstrates a bi-directional relationship (i.e., a covariance);
5. Covariance arrows should only be allowed for exogenous variables; and
6. For every endogenous variable, a residual term should be added in the model. Generally, a residual term is a circle with the letter E written in it, which stands for error.

Model Estimation

In SEM, the parameters of a proposed model are estimated by minimizing the discrepancy between the empirical (sample) covariance matrix, S, and a covariance matrix, Σ, implied by the model (population). When elements in the matrix S minus the elements in the matrix Σ equal zero $(S - \Sigma, = 0)$, then one has a perfect model fit to the data. The model estimation process uses a *fitting function or estimation procedures* to minimize the difference between Σ and S. Several fitting functions are available. In AMOS, the following estimation procedures are available: *unweighted or ordinary least squares* (ULS or OLS), *generalized least squares* (GLS), *asymptotically distribution free* (ADF), *scale-free least squares* (SFLS), and *maximum likelihood* (ML).

The least-squares criterion minimizes the sum of squared residuals between the observed and predicted values of y. In the regression setting, this criterion is only optimal if the assumption of homoscedasticity is satisfied. When this assumption is violated, weighted least-squares (WLS) regression can be used instead, which minimizes a weighted sum of squares, with the weights reflecting the different variances of individual elements. In SEM, the assumption of homogeneity is never plausible, because in the place of y we have the very different elements of the sample covariance matrix, whose variances have no reason to be the same (these "variances of the variances," or fourth-order moments, are related to each variable's kurtosis). Moreover, while in regression we assume that the observations are independent of each other, in SEM the elements of the sample covariance matrix are not in fact independent, and additional weights related to their covariances also need to be estimated. Thus, in SEM, LS estimation is rarely the optimal choice.

GLS and ADF, also referred to in the literature as AGLS, arbitrary distribution generalized least squares, differ in the assumptions the researcher must make about the data and in the choice of weights. GLS is appropriate when the variables have no excess kurtosis, so that the weights are greatly simplified. This estimator is appropriate when the data are normally distributed, for example. ADF does not require any assumptions and estimates all the weights from the data before using them in a fitting function. Because estimating these weights accurately requires large samples, this method almost never works well unless the sample size is very large (perhaps a thousand or more) or the model is very simple.

In *scale-free least squares estimation* (**SFLS**), the minimum of the fitting function is independent of the scale of the variables. For example, with estimation that is not scale-free, the solutions based on the covariance matrix input and the correlation matrix input can differ. The GLS and ML estimation methods also are sale free, which means that if we transform the scale of one or more of our observed variables, the untransformed and the transformed variables will yield estimates that are properly related, that is, that differ by the transformation.

Maximum likelihood (**ML**) estimates model parameters that have the greatest chance of reproducing the observed data. First detailed by Fisher in 1920, ML proposes the estimation of a parameter by that value for which the likelihood function is at a maximum for the given data. Essentially, the principle is that the value of the parameter under which the obtained data could have had highest probability of arising must be intuitively our best estimator of the population value. For example, suppose a researcher seeks to identify the best estimate of the population mean given a particular sample mean. The sample mean is the ML estimator for the population mean. That is, the researcher chooses from among several estimates of the population mean by selecting the estimate that maximizes the likelihood that the discrepancy between that estimated population mean and the observed sample information must be attributed to sampling error.

In SEM, the statistical criterion minimized in ML estimation is related to discrepancies between the observed covariances and those predicted by the researcher's model. The mathematics of how ML estimation actually goes about generating a set of parameter estimates that minimize its fitting function are complex, and it is beyond the scope of this section to describe them in detail. An accessible presentation is available in Nunnally & Bernstein (1994).

Because MLE has no algebraic formulas similar to the equations used in OLS regression, its solution requires a computer capable of examining many parameter sets until the best choice is identified. The procedure begins with an initial estimate. A series of iterations, or cycles, produces new estimates and compares them to the previous ones. The iterations continue until successive estimates differ from the preceding ones by less than a specified small amount. Computer programs usually issue a warning message if iterative estimation is unsuccessful. When this occurs, whatever final set of estimates was derived by the computer may warrant little confidence. A tactic is to increase the probability of convergence is to change the program's default limit on the number of iterations to a higher value (e.g., from 50 to 100). Allowing the computer more "tries" may lead to a converged solution.

When a raw data file is analyzed, standard ML estimation assumes there are no missing values. For large samples, MLE parameter estimates are unbiased, efficient, and normally distributed, and thus allow significance tests. More specifically, ML estimation assumes that the population distribution for the endogenous variables is multivariate normal. A growing body of research suggest that although the values of parameter estimates generated by ML are relatively robust against nonnormality, results of statistical tests may be positively biased, which means that they lead to the rejection of the null hypothesis too often. (c.f. Satorra & Bentler, 1994; Hoyle & Panter, 1995). Because ML is so widely available and is the most widely researched estimator among those otherwise available, it has been suggested that the use of an estimation method other than ML requires explicit justification (Hoyle & Panter, 1995). If characteristics of the data raise question as to the appropriateness of ML, then the results of alternative estimation procedures might be reported in summary form if they contradict ML results or in a footnote if they corroborate them. Alternatively, researchers may consider applying linear transformations, such as square root and logarithmic, which are discussed in chapter 1.

However complex the mathematics of ML estimation, the interpretation of ML estimates for path models is relatively straightforward. Path coefficients are interpreted as regression coefficients in multiple regression, which means that they control for correlations among multiple presumed causes. In standard ML estimation, standard errors are calculated only for the unstandardized solution. This means that results of statistical tests (i.e., ratios of parameter estimates over their standard errors) are available only for the unstandardized solution.

Also, the level of statistical significance of an unstandardized estimate does not automatically apply to its standardized counterpart. Furthermore, standard ML estimation may derive incorrect standard errors when the variables are standardized. However, there are basically two ways to obtain more accurate estimated standard errors for the standardized solution. Some SEM computer programs, such as AMOS, use bootstrapping to generate standard errors for standardized estimates. Another method is constrained estimation, which is available as a user-specified option in some SEM computer programs including SEPATH and RAMONA.

Model Fit and Interpretation

Finding a statistically significant theoretical model that also has practical and substantive meaning is the primary goal of using SEM to test theories. Consequently, once estimated, the model's fit to the data must be evaluated. A *model's fit* refers to its ability to reproduce the data (i.e., usually the variance–covariance matrix). The objective is to determine whether the associations among measured and latent variables in the researcher's estimated model adequately reflect the observed associations in the data. The perspective commonly taken by social workers and other applied researchers is that approximating observed data is acceptable and can result in important contributions to the literature. It should be noted that a good-fitting model is not necessarily a valid model. Accordingly, researchers also should consider the implications of accepting one explanatory model over another in terms of theory-related criteria. More specifically, researchers should evaluate fit in terms of (1) statistical significance and strength of estimated parameters; (2) variance accounted for in endogenous observed and latent variables; and (3) how well the overall model fits the observed data, as indicated by a variety of fit indices.

Multiple indices are available to evaluate model fit. See Figure 5.4 for a summary of commonly reported fit indices, including generally accepted thresholds. These fit indices reported by most software programs, including AMOS. SEM software packages, such as AMOS, produce a large number of alternative measures of model fit. These additional measures are discussed in the context of the annotated example presented below. See Kaplan (2009) or Mulaik (2009) for more information about other fit statistics in SEM.

Following Kline (2011), Boomsma (2000), MacCallum and Austin (2000), and McDonald and Ho (2002), the following fit indexes are reviewed here: (1) model chi-square, (2) goodness of fit Index (GFI), (3) comparative fit index (CFI), (4) root mean square error of approximation (RMSEA), and (5) standardized root mean square residual (SRMR).

The chi-square (χ^2) test is a conventional null hypothesis significance test (NHST). According to Kline (2011), as the most basic measure of model fit, the χ^2 test is the product $(N-1)$ F_{ML}, where F_{ML} is the value of the statistical criterion (fit function) minimized in ML estimation and $(N-1)$ is one less than the sample size. In large samples and assuming multivariate normality, the product $(N-1)$ F_{ML} follows a central chi-square distribution with degrees of freedom equal to that of the researcher's model, or df_M. This statistic is also known as the **likelihood ratio chi-square** or **generalized likelihood ratio**.

The null hypothesis (H_0) is that the postulated model holds in the population; that is, the sample covariance matrix equals the population covariance matrix. AMOS reports the value of chi-square as **CMIN**. In contrast to traditional significance testing, the researcher prefers a nonsignificant chi-square, since this result indicates that the predicted model is congruent with the observed data. The higher the probability level (p-value) associated with chi square, the better the fit. A significant chi-square indicates lack of satisfactory model fit.

Index	Shorthand	Acceptable Fit
Chi-square (CMIN)	χ^2 test	Ratio of χ^2 to $df \leq 3$
Root Mean Square Residual	RMR	$\leq .05$
Goodness of Fit Index	GFI	$\geq .95$
Adjusted Goodness of Fit Index	AGFI	$\geq .95$
Parsimony Goodness of Fit Index	PGFI	$\geq .95$
Normed Fit Index	NFI	$\geq .95$
Relative Fit Index	RFI	$\geq .95$
Incremental Fit Index	IFI	$\geq .95$
Tucker-Lewis Index	TLI	$\geq .95$
Comparative Fit Index	CFI	$\geq .95$
Parsimony Normed Fit Index	PNFI	Larger values indicate better fit
Parsimony Comparative Fit Index	PCFI	Larger values indicate better fit
Root Mean Square Error of the Approximation	RMSEA	$\leq .05$
Akaike Information Criterion	AIC	Smaller values indicate better fit
Browne-Cudeck Criterion	BCC	Smaller values indicate better fit
Bayes Information Criterion	BIC	Smaller values indicate better fit
Consistent AIC Criterion	CAIC	Smaller values indicate better fit
Expected Cross-Validation Index	ECVI	Smaller values indicate better fit
Modified Expected Cross-Validation Index	MECVI	Smaller values indicate better fit

Figure 5.4 Commonly Reported for Fit Indices.

The chi-square (χ^2) test has at least two limitations. First, the chi-square test offers only a dichotomous decision strategy implied by a statistical decision rule and cannot be used to quantify the degree of fit along a continuum with some prespecified boundary. Second, as with most statistics, large sample sizes increase power, resulting in significance with small effect sizes. Consequently, a nonsignificant χ^2 may be unlikely, although the model may be a close fit to the observed data. Despite these limitations, researchers almost universally report the χ^2 (Martens, 2005).

The **Goodness-of-Fit Index (GFI)** estimates the proportion of covariances in the sample data matrix explained by the model. That is, the GFI estimates how much better the researcher's model fits compared with no model at all (Jöreskog, 2004). The range of values for this pair of approximate fit indexes is generally 0–1.0, where 1.0 indicates the best fit. A general formula is

$$GFI = 1 - \frac{C_{res}}{C_{tot}} \qquad (5.1)$$

where C_{res} and C_{tot} estimate, respectively, the residual and total variability in the sample covariance matrix. The numerator in the right side of Equation 5.1 is related to the sum of the squared covariance residuals, and the denominator is related to the total sum of squares in the data matrix. GFI should by equal to or greater than .90 to indicate good fit. GFI is less than or equal to 1. A value of 1 indicates a perfect fit. GFI tends to be larger as sample size increases. GFI >.95 indicates good fit. GFI index is roughly analogous to the multiple R-square in multiple regression because it represents the overall amount of the covariation among the observed variables that can be accounted for by the hypothesized model. One limitation of the GFI is that its expected values vary with sample size. Another limitation is that values of the GFI sometimes fall outside of the range 0 – 1.

The **Comparative Fit Index** (*CFI*) (Bentler, 1990) measures the relative improvement in the fit of the researcher's model over that of a baseline model, typically the ***independence or null model***, which specifies no relationships among variables. CFI ranges from 0 to 1.0. Values close to 1 indicates a very good fit, >.9 or close to .95 indicates good fit, by convention, CFI should be equal to or greater than .90 to accept the model. The formula is

$$CFI = 1 - \frac{\chi^2_M - df_M}{\chi^2_B - df_B} \qquad (5.2)$$

where the numerator and the denominator of the expression on the right side of the equation estimates the chi square noncentrality parameter for, respectively, the researcher's model and the baseline model. Note that noncentrality parameters reflect the extent to which the null hypothesis is false. For example, the traditional χ^2 test assumes that the null hypothesis is true ($\chi^2 = 0$) in the population. This test relies on the "central" distribution of χ^2 values. Because the researcher is hoping *not* to reject the null hypothesis, it is argued that it is more appropriate to test the alternative hypothesis (H_a). This test of H_a would rely on a "noncentral" chi-square distribution that assumes H_a is true in the population. This approach to model fit uses a chi-square equal to the *df* for the model as having a perfect fit (as opposed to $\chi^2 = 0$). Thus, the noncentrality parameter estimate is calculated by subtracting the *df* of the model from the chi-square ($\chi^2 - df$).

CFI is relatively insensitive to sample size (Fan, Thompson, and Wang, 1999). However, one limitation is that the null hypothesis that the baseline model is better than the independence model is almost always true. This is because the assumption of zero covariances among the observed variables is improbable in most studies. Although it is possible to specify a different, more plausible baseline model—such as one that allows the exogenous variables only to covary—and compute by hand the value of an incremental fit index with its equation, this is rarely done in practice. Widaman and Thompson (2003) describe how to specify more plausible baseline models.

The root-mean-square error of the approximation (RMSEA) is another index based that is based on the noncentrality parameter (Steiger, 1990). The formula is

$$\text{RMSEA} = \sqrt{\frac{\chi_M^2 - df_M}{df_M(N-1)}} \tag{5.3}$$

The numerator is the noncentrality parameter, and the denominator is the product of the model degrees of freedom and one less than the sample size. That is, the value of the RMSEA decreases as there are more degrees of freedom (greater parsimony) or a larger sample size, keeping all else constant. However, the RMSEA does not necessarily favor models with more degrees of freedom. This is because the effect of the **correction for parsimony** diminishes as the sample size becomes increasingly large (Kline, 2011).

Hu and Bentler (1995) suggest values below .06 indicate good fit. The RMSEA values are classified into four categories: close fit (.00 – .05), fair fit (.05 – .08), mediocre fit (.08 – .10), and poor fit (over .10). RMSEA tends to improve as we add variables to the model, especially with larger sample size. A confidence interval can be computed for this index. According to Kline (2011), the population parameter estimated by the RMSEA is often designated as ε. Because the RMSEA's distribution values are known, a confidence interval around the point estimate of the RMSEA can be constructed to indicate the level of its precision. In computer output, the lower and upper bounds of the 90% confidence interval for ε are often printed along with the sample value of the RMSEA, the point estimate of ε. Using this confidence interval, evaluating the null hypothesis can be examined more precisely. In using these confidence intervals, a null hypothesis could be rejected in favor of accepting the alternative if the entire range of the confidence interval is less than .05. So, like the χ^2 statistic, it is possible to use RMSEA to evaluate the null hypothesis that a model fits the data exactly. According to Kline (2011), the width of this confidence interval is generally larger in smaller samples, which indicates less precision. The bounds of the confidence interval for ε may not be symmetrical around the sample value of the RMSEA, and, ideally, the lower bound equals zero. If the upper bound of the confidence interval for ε exceeds a value that may indicate "poor fit," then the model warrants less confidence.

Unlike the chi-squared statistic, RMSEA is less affected by sample size problems. Relatively little information is currently available on the performance of RMSEA when data are nonnormal. However, available information suggests that RMSEA may perform less optimally when there are large sample sizes and relatively small degrees of freedom (Quintana & Maxwell, 1999). The interpretation of the RMSEA and the lower and upper bounds of its confidence interval depends on the assumption that this statistic follows noncentral chi-square distributions. Evidence suggests that the aforementioned assumption may not be tenable that (cf. Olsson, Foss, & Breivik, 2004; Yuan, 2005). RMSEA tends to impose a harsher penalty for complexity on smaller models with relatively few variables or factors. This is because smaller models may have relatively few degrees of freedom, but larger models may have more "room" for higher df_M values. Consequently, the RMSEA may favor larger models (Breivik & Olsson, 2001).

The standardized root-mean-square residual (SRMR) is based on transforming both the sample covariance matrix and the predicted covariance matrix into correlation matrices. The SRMR is thus a measure of the mean absolute correlation residual, the overall difference between the observed and predicted correlations. The Hu and Bentler (1999) threshold of SRMR ≤ .08 for acceptable fit was not a very demanding standard. This is because if the average absolute correlation residual is around .08, then many individual values could exceed this value, which would indicate poor explanatory power at the level of pairs of observed variables. It is better to actually inspect the matrix of correlation residuals and describe their pattern as part of a diagnostic assessment of fit than just to report the summary statistic SRMR.

In addition to considering overall model fit, it is important to consider the significance of estimated parameters, which are analogous to regression coefficients. As with regression, a model that fits the data quite well but has few significant parameters would be meaningless. At a minimum, the researcher should inspect model estimates to determine if proposed parameters were significant and in the expected direction.

Model Modification

Rarely is a proposed model the best-fitting model. Consequently, modification, also termed respecification, may be needed. This involves adjusting the estimated model by freeing (estimating) or setting (not estimating) parameters. Post hoc modifications of the model are often based on *modification indices*. Improvement in fit is measured by a reduction in chi-square, which makes the chi-square fit index less likely to be found significant (recall a finding of significance corresponds to rejecting the model as one that fits the data). For each fixed and constrained parameter (coefficient), the modification index reflects the predicted decrease in chi-square if a single fixed parameter or equality constraint is removed from the model by eliminating its path, and the model is re-estimated. One arbitrary rule of thumb is to consider eliminating paths associated with parameters whose modification index exceeds 10. However, another approach is to eliminate the parameter with the largest MI, then see the effect as measured by the chi-square fit index.

Modification is a controversial topic, which has been likened to the debate about post hoc comparisons in ANOVA (MacCallum & Austin, 2000; McDonald & Ho, 2002). The suggested modifications, however, may or may not be supported on theoretical grounds. As with ANOVA and regression, problems with model modification include capitalization on chance and results that are specific to a sample because they are data driven. Although there is disagreement regarding the acceptability of post hoc model modification, statisticians and applied researchers alike emphasize the need to clearly state when there was post hoc modification rather than imply that analyses were *a priori.*

Researchers are urged not to make too many changes based on modification indices, even if such modifications seem sensible on theoretical grounds. Note that SEM takes a confirmatory approach to model testing; one does not try to find the best model or theory via data using SEM. Rather than data-driven post hoc modifications (which may be very inconsistent over repeated samples), it is often more defensible to consider multiple alternative models *a priori.* That is, multiple models (e.g., based on competing theories or different sides of an argument) should be specified prior to model fitting, and the best-fitting model should be selected among the alternatives. Because a more complex model, assuming it is identified, will generally produce better fit, and different models can produce the same fit, theory is imperative in model testing.

In conclusion, it is worth noting that although SEM allows the testing of causal hypotheses, a well-fitting SEM model does not and cannot prove causal relations without satisfying the necessary conditions for causal inference (e.g., time precedence, robust relationship in the presence or absence of other variables). A selected well-fitting model in SEM is like a retained null hypothesis in conventional hypothesis testing; it remains plausible among perhaps many other models that are not tested but may produce the same or better level of fit. SEM users are cautioned not to make unwarranted causal claims. Replications of findings with independent samples are recommended, especially if the models are obtained with post hoc modifications.

STRENGTHS AND LIMITATIONS OF SEM

Strengths and Limitations of SEM

SEM has the following analytical strengths:

1. Enables researchers to answer a set of interrelated research questions by modeling the relationships among multiple IVs and DVs simultaneously. That is, SEM permits complicated

variable relationships to be expressed through hierarchical or nonhierarchical, recursive or nonrecursive structural equations;

2. Allows researchers to specify latent variable models that provide separate estimates of relations among latent constructs and their manifest indicators (the measurement model) and of the relations among constructs (the structural model);

3. Availability of measures of global fit that can provide a summary evaluation of even complex models that involve a large number of linear equations. Most alternative procedures that might be used in place of SEM (e.g., MANOVA and MANCOVA) to test such models would provide only separate "minitests" of model components that are conducted on an equation-by-equation basis; and

4. Allows researchers to directly test the model of interest, also termed the theoretical or alternative hypothesis, directly. In most statistical contexts, the researcher's theoretical hypothesis is aligned with the alternative hypothesis rather than the null hypothesis. In contrast, in SEM the theoretical hypothesis is often aligned with the null hypothesis, which specifies that the model fits exactly or at least approximately (e.g., MacCallum et al. 1993). It should be noted that between-group comparisons of factor means, for examples, represents an exceptions to this conclusion;

SEM has the following analytical limitations:

1. A well-fitting SEM model does not demonstrate causality. That is, a well-fitting model in SEM is plausible among other models that are not tested but may produce the same or better level of fit;

2. Cannot be used as an exploratory modeling strategy. Theory should be used to guide model specification;

3. Like all statistical strategies, SEMs are approximations of reality. For example, important variables may be omitted. Such omissions present a misleading picture of the measurement.

Although the problem of omitted variables is not unique to SEM, the implications of model misspecification in SEM must be recognized. According to Tomarken and Waller (2005), users often underestimate the importance of residual variance and covariance terms for generating a model with acceptable fit. More specifically, if residual variance and covariance terms were to be omitted, at least some models would fit poorly. Tomarken and Waller (2005) note a paradox: "the residual parameterizations afforded by SEM software can mask the limitations of a rather incomplete model" (p. 49); and

4. Measures of global fit do not directly test lower-order components of a model, such as path coefficients and relevant quantities that can be derived from such parameters, such as the proportion of variance in an endogenous variable that is accounted for by the specified predictors in the model. Researchers must be aware that a model can fit perfectly yet be associated with problematic lower-order components (e.g., parameter estimates that are biased, small in magnitude or opposite to theoretical expectations).

ANNOTATED EXAMPLE

A researcher plans to examine the relationship between factors that influence postadoption service utilization and positive adoption outcomes. Specifically, the study tests a model that links (1) factors influencing the utilization of postadoption services (*parents' perceptions of self-efficacy, relationship satisfaction between parents,* and *attitudes toward adoption*) with (2) *service utilization,* and (3) positive adoption outcomes (*satisfaction with parenting* and *satisfaction with adoption agency*). See Figure 5.5. Note that latent variables are not included in the current model, since

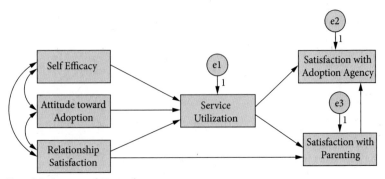

Figure 5.5 Annotated Example.

instruments are unidimensional with a high reliability. Unidimensionality and reliability of instruments is discussed further in chapter 2 (Annotated Example section) and in chapter 6.

Determining the Minimally Sufficient Sample Size

As discussed earlier in this chapter, structural equation modeling (SEM) does not use raw data. Instead, in SEM the variance (or covariance) matrix is used. The number of observations in SEM is defined as the number of covariances in the matrix rather than the number of cases in a data set. The number of observations or covariances in SEM can be calculated as follows:

$$v(v + 1)/2,$$

where v is the number of variables in the model. In the current model, there are seven variables (i.e., *self-efficacy, satisfaction with marital relationship, attitude toward adoption, knowledge of agency services, service utilization, satisfaction with parenting,* and *satisfaction with adoption agency*). Therefore, the number of variables in this example is $6(6 + 1)/2$, or 21.

The researcher determines sample size by following the approach proposed by MacCallum and Browne (1993), which was described earlier in this chapter. This approach uses the RMSEA (Brown, 2006; Hu & Bentler, 1999; Steiger, 1990). This index weighs absolute fit, which declines whenever a parameter is removed from the model, against model complexity, such that the benefits of parsimony are considered along with fit (Steiger, 1990). Models fitting with RMSEA <.05 are usually considered as having a "close" fit, .05 to .08 as having a "fair" fit, .08 to .10 as having a "mediocre" fit, and above .10 as having a "poor" fit (MacCallum et al., 1993).

A program called NIESEM performs power analysis according to the approach proposed by MacCallum and Browne, (1993). NIESEM is free and available for download from http://rubens.its.unimelb.edu.au/~dudgeon/. To estimate sample size with NIESEM, the researcher proceeds as follows:

1. Selects Power calculations
2. Selects estimate N for given power
3. Inputs power equals 0.80

4. Inputs the null hypothesized RMSEA value equals 0.00
5. Inputs the alternative hypothesized RMSEA value equals 0.05
6. Inputs the chosen alpha level equals 0.05
7. Inputs the degrees of freedom of the model equals 6(6 + 1)/2 = 21
8. Inputs the number of groups in the model equals 1
9. Clicks OK

The output is as follows (see Figure 5.6) and the estimated sample size equals 408.

An alternative strategy for estimating sample size for SEM is provided by the following webpage: http://timo.gnambs.at/en/scripts/power-forsem. This webpage generates syntax for SPSS and R to estimate sample size for various measures of model fit, including RMSEA, GFI, AGFI. This webpage http://www.datavis.ca/sasmac/csmpower.html provides an SAS macro, csmpower, to calculate retrospective or prospective power computations for SEM using the method of MacCallum and Browne (1993). Their approach allows for testing a null hypothesis of "not-good-fit," so that a significant result provides support for good fit. Effect size in this approach is defined in terms of a null hypothesis and alternative hypothesis value of the root-mean-square error of approximation (RMSEA) index. These values, together with the degrees of freedom (df) for the model being fitted, the sample size (n), and error rate (alpha), allow power to be calculated.

```
----- CSM Power Analysis ----

RMSEA Null Value =            0.000

RMSEA Alternative Value = 0.050

Alpha significance level= 0.050

Degrees of freedom =            21

Number of groups =             1


Desired power =            0.800

----------------------------
Estimated sample size =    408
----------------------------
```

Figure 5.6 CSM Power.

To perform the SEM analysis in the researcher proceeds as follows:

1. Selects the data set that will be used to test the model by clicking File Data Files File Name to browse to and select the file;
2. Draws the model diagram by using the Diagram dropdown menu (see Figure 5.7);
3. Draws observed variables by selecting Draw Observed, then uses the cursor to draw the five observed variables (rectangles) in the model;
4. Names an observed variable by right-clicking on a rectangle, and under text tab, adds the variable name;
5. Draws latent variables (not included in the current mode) by selecting Draw Observed, and then uses the cursor to draw the five observed variables (rectangles) in the model;
6. Names a latent variable by right-clicking on a rectangle, and under text tab, adding the variable name;
7. Draws paths by selecting Draw Path, and using the cursor;
8. Draws error terms by selecting Draw Unique Variable and using the cursor; and
9. Names an error term by right-clicking a rectangle, and under text tab adding the error term's name.

Once the model is illustrated,

10. Clicks View Analysis Properties Estimation tab;
11. Selects Maximum likelihood;
12. Selects Fit saturated and independence models;
13. Clicks Output tab;
14. Selects Standardized estimates;
15. Selects Residual moments;
16. Selects Modification indices;
17. Selects Indirect, direct, and total effects;
18. Selects Covariances of estimates;
19. Selects Correlations of estimates;
20. Selects Tests for normality and outliers;
21. Once the model is illustrated and analysis properties are selected, clicks Analyze, Calculate Estimates; and

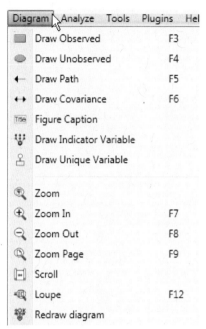

Figure 5.7 Diagram Dropdown Menu.

22. To view results clicks View Text Output.

(Note that the toolbar icons may also be used to view analysis properties, calculate estimates, and view output.)

AMOS next prints out a large number of alternative measures of model fit (see Figure 5.3). Each measure is calculated for three models. "Default model" is the "current model" as specified by the researcher. The "independence model" is the model in which variables are assumed to be uncorrelated with the dependent(s), so if the fit for "your model" is no better than for the "independence model," then your model should be rejected. The "saturated model" is one with no constraints and will always fit any data perfectly, so normally your model will have a measure of fit between the saturated and independence models.

The output is as follows: NPAR is the number of parameters being estimated in the model and is not a measure of fit. AMOS reports the value of the χ^2 test as **CMIN,** and its associated p-value as **P(CMIN).** If P(CMIN) is less than .05, we reject null hypothesis that the data are a perfect fit to the model. By this criterion the present model is not rejected. Recall that nonsignificant chi-square (e.g., $p > .05$) indicates that the parameters that were estimated for the model fit the data (Please see Figure 5.8).

CMIN/DF is the χ^2 *test* divided by the current model's degrees of freedom (*df*). Some researchers allow values as large as 5 as being an adequate fit, but conservative use calls for rejecting models with relative chi-square greater than 3. By this criterion the current model is not rejected (Please see Figure 5.8).

One important problem with chi-square is that with large samples significance is easy to obtain. Given that large sample are recommended for the SEM technique, this presents a dilemma. A solution has been to develop what are called fit indexes which are based on the chi-square but which control in some way for sample size.

RMR is the Root-Mean-Square Residual, which is the square root of the mean squared amount by which the sample covariances differ from the estimated covariances, estimated on the assumption that your model is correct; the smaller the RMR, the better the fit. An RMR of zero indicates a perfect fit. The closer the RMR to 0 for a model being tested, the better the model fit, and an RMR value smaller than .05 suggests good fit. For these data, RMR equals 9.370, which suggests that the current model is not a good fit (Please see Figure 5.9).

GFI is the Goodness of Fit Index. GFI varies from 0 to 1, but theoretically can yield meaningless negative values. By convention, GFI should by equal to or greater than .90 to accept the model. By this criterion the present model (GFI = .979) is accepted (Please see Figure 5.9).

AGFI is the Adjusted Goodness of Fit Index. AGFI is a variant of GFI. AGFI also varies from 0 to 1, but theoretically can yield meaningless negative values. AGFI should also be at least .90. By this criterion the present model (AGFI = .913) is accepted (Please see Figure 5.9).

Model	NPAR	CMIN	DF	P(CMIN)	CMIN/DF
Default model (current model)	16	6.499	5	.261	1.300
Saturated model	21	.000	0		
Independence model	6	87.629	15	.000	5.842

Figure 5.8 AMOS *CMIN* Output.

PGFI is the Parsimony Goodness of Fit Index. It is a variant of GFI that penalizes GFI by multiplying it times the ratio formed by the degrees of freedom in your model and degrees of freedom in the independence model. The PGFI ranges from 0 to 1, with higher values indicating a more parsimonious fit. There is no standard for how high the index should be to indicate parsimonious fit. This index is best used to compare two competing theoretical models. That is, calculate PGFI for each model and choose the model with the highest level of parsimonious fit (Please see Figure 5.9).

NFI is the Normed Fit Index, which varies from 0 to 1, with 1 = perfect fit. By convention, NFI values below .90 indicate a need to respecify the model. By this criterion the present model (NFI = .926) is accepted (Please see Figure 5.10).

RFI is the Relative Fit Index, which is not guaranteed to vary from 0 to 1. RFI close to 1 indicates a good fit. By this criterion the present model (RFI = .778) is accepted (Please see Figure 5.10).

IFI is the Incremental Fit Index, which is not guaranteed to vary from 0 to 1. IFI close to 1 indicates a good fit and values above .90 an acceptable fit. By this criterion the present model (IFI = .982) is accepted (Please see Figure 5.10).

TLI is the Tucker–Lewis Index, also called the Bentler–Bonett non-normed fit index (NNFI). TLI is not guaranteed to vary from 0 to 1. TLI close to 1 indicates a good fit. By this criterion the present model (TLI = .938) is accepted (Please see Figure 5.10).

CFI is the Comparative Fit Index, which varies from 0 to 1. CFI close to 1 indicates a very good fit, and values above .90 an acceptable fit. By this criterion the present model (CFI = .979) is accepted (Please see Figure 5.10).

Note that the **DELTA** and **RHO** headings are alternative names for these measures.

Model	RMR	GFI	AGFI	PGFI
Default model (current model)	9.370	.979	.913	.233
Saturated model	.000	1.000		
Independence model	53.318	.772	.681	.552

Figure 5.9 RMR, GFI, AGFI, PGFI Output

PRATIO is the Parsimony Ratio, which is the ratio of the degrees of freedom in your model to degrees of freedom in the independence (null) model. PRATIO is not itself a goodness-of-fit test, but is used in goodness-of-fit measures like PNFI and PCFI that reward parsimonious models with relatively few parameters to estimate in relation to the number of variables and relationships in the model (Please see Figure 5.11).

PNFI is the Parsimony Normed Fit Index, equal to the PRATIO times NFI. There is no standard for how high the index should be to indicate parsimonious fit. This index is best used to compare two competing theoretical models (Please see Figure 5.11).

PCFI is the Parsimony Comparative Fit Index, equal to PRATIO times CFI. There is no standard for how high the index should be to indicate parsimonious fit. This index is best used to compare two competing theoretical models (Please see Figure 5.11).

NCP is the noncentrality parameter. It and FO are used in the computation **of RMSEA**, the root-mean-square error of approximation, which incorporates the discrepancy function criterion (comparing observed and predicted covariance matrices) and the parsimony criterion (see above). For each, LO 90 and HI 90 indicate 90% confidence limits on the coefficient (Please see Figure 5.12).

FMIN is the minimum fit function. It can be used as an alternative to CMIN to compute CFI, NFI, NNFI, IFI, and other fit measures (Please see Figure 5.13).

RMSEA is the Root-Mean-Square Error of the Approximation. The RMSEA values usually are classified into four categories: close fit (.00 – .05), fair fit (.05 – .08), mediocre fit (.08 – .10), and poor fit (over .10). By convention, there is good model fit if RMSEA less than or equal to .05. For these data, RMSEA equals .055, which suggests that the model is a fair fit. PCLOSE tests the null hypothesis that RMSEA is no greater than .05. Since PCLOSE equals .394, we do

Model	NFI Delta1	RFI rho1	IFI Delta2	TLI rho2	CFI
Default model	.926	.778	.982	.938	.979
Saturated model	1.000		1.000		1.000
Independence model	.000	.000	.000	.000	.000

Figure 5.10 NFI, RFI, IFI, TLI, CFI Output.

Model	PRATIO	PNFI	PCFI
Default model (current model)	.333	.309	.326
Saturated model	.000	.000	.000
Independence model	1.000	.000	.000

Figure 5.11 PRATIO, PHFI, PCFI Output.

not reject the null hypothesis and conclude that RMSEA is no greater than .05 (Please see Figure 5.14).

AIC is the Akaike Information Criterion. The AIC measure indicates a better fit when it is smaller. The measure is not standardized and is not interpreted for a given model. For two models estimated from the same data set, the model with the smaller AIC is to be preferred. The AIC makes the researcher pay a "penalty" for every parameter that is estimated. The absolute value of AIC has relatively little meaning; rather the focus is on the relative size, the model with the smaller AIC being preferred (Please see Figure 5.15).

BCC is the Browne–Cudeck Criterion, also called the Cudeck & Browne single-sample cross-validation index. The BCC should be close to .9 to conclude a model is a good fit. BCC penalizes for model complexity (lack of parsimony) more than AIC. For two models estimated from the same data set, the model with the smaller BCC is to be preferred (Please see Figure 5.15).

BIC is the Bayes Information Criterion, also known as Akaike's Bayesian information criterion (**ABIC**). BIC penalizes for sample size as well as model complexity. Specifically, BIC penalizes for additional model parameters more severely than does AIC. For two models estimated from

Model	NCP	LO 90	HI 90
Default model	1.499	.000	12.357
Saturated model	.000	.000	.000
Independence model	72.629	46.848	105.919

Figure 5.12 NCP Output.

Model	FMIN	F0	LO 90	HI 90
Default model	.066	.015	.000	.125
Saturated model	.000	.000	.000	.000
Independence model	.885	.734	.473	1.070

Figure 5.13 FMIN Output.

the same data set, the model with the smaller BIC is to be preferred (Please see Figure 5.15).

CAIC is the Consistent AIC Criterion, which also penalizes for sample size as well as model complexity (lack of parsimony). The penalty is greater than AIC or BCC but less than BIC. For two models estimated from the same data set, the model with the smaller CAIC is to be preferred (Please see Figure 5.15).

ECVI is the Expected Cross-Validation Index. It is another variant on AIC. For two models estimated from the same data set, the model with the smaller ECVI is to be preferred. **MECVI** is the Modified Expected Cross-Validation Index. It is a variant on BCC. *MECVI* is a variant on BCC, differing in scale factor. Compared to ECVI, a greater penalty is imposed for model complexity. Lower is better between models. For two models estimated from the same data set, the model with the smaller MECVI is to be preferred (Please see Figure 5.16).

Hoelter's critical *N* or **Hoetler index** is the largest sample size at which the researcher would accept the model at the .05 or .01 levels. This offers a perspective on the chi-square fit index, which has the problem that the larger the sample size, the more likely the rejection of the model and the more likely a Type II error. In this case, actual sample size was 408 and the model was accepted. If the sample size had been only 164, it would have been accepted at the .05 level (Please see Figure 5.17).

Model	RMSEA	LO 90	HI 90	PCLOSE
Default model	.055	.000	.158	.394
Independence model	.221	.178	.267	.000

Figure 5.14 RMSEA Output.

Model	AIC	BCC	BIC	CAIC
Default model	38.499	40.934	80.182	96.182
Saturated model	42.000	45.196	96.709	117.709
Independence model	99.629	100.542	115.260	121.260

Figure 5.15 AIC, BCC, BIC, CAIC Output.

To calculate the **Standardized Root-Mean-Square Residual** (*SRMR*) in AMOS,

1. Select Analyze Calculate Estimates;
2. Select Plugins Standardized RMR, and a blank Standardized RMR dialog box is displayed; and
3. Re-select Analyze Calculate Estimates, and the Standardized RMR dialog will display SRMR.

For these data, SRMR equals .0513. Therefore, the model is a fair to good fit. SRMR ≤ .08 suggests good fit. Additionally, the matrix of correlation was inspected. No unusual coefficients were observed. No modification index was greater than zero, and, consequently, no model modifications were suggested.

In addition to considering overall model fit, it is important to consider the significance of estimated parameters, which are analogous to regression coefficients. As with regression, a model that fits the data quite well but has few significant parameters would be meaningless. At

Model	ECVI	LO 90	HI 90	MECVI
Default model	.389	.374	.499	.413
Saturated model	.424	.424	.424	.457
Independence model	1.006	.746	1.343	1.016

Figure 5.16 ECVI, MECVI Output.

Model	HOELTER .05	HOELTER .01
Default model	469	431
Independence model	329	335

Figure 5.17 Hoelter's Critical N Output.

a minimum, the researcher should inspect model estimates to determine if proposed parameters were significant and in the expected direction.

For the final model (see Figure 5.18), correlation coefficients for exogenous variables, and unstandardized and standardized regression coefficients for endogenous variables were calculated. Of note for the exogenous variables, *Self Efficacy* and *Relationship satisfaction* were moderately correlated ($r = .47$).

Of note for the endogenous variables, the standardized coefficients between *Relationship Satisfaction* and *Service Utilization* was −.52, *Self Efficacy* and *Service Utilization* was −.15, *Relationship Satisfaction* and *Satisfaction with Parenting* was .16, *Service Utilization* and *Satisfaction with Parenting* was .19, and *Satisfaction with Parenting* and *Satisfaction with Adoption Agency* was .28. That is, the best predictor of *Satisfaction with Parenting* is *Service Utilization*, and the best predictor of *Service Utilization* is *Sat with Parenting*.

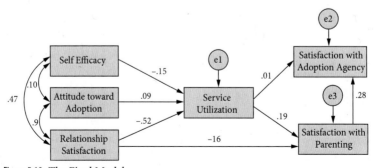

Figure 5.18 The Final Model.

REPORTING THE RESULTS OF A SEM

1. Sample selection strategy and size with justifications (e.g., power analysis);
2. Extent and handling of issues related to missing data, normality, outliers, and multicollinearity;
3. Model identification;
4. Software program used;
5. Estimation method;
6. Model fit;
7. Model coefficients;
8. Residual analysis (i.e., predicted and actual covariance matrix examination);
9. Model modification, including rationale; and
10. Diagram of final model.

RESULTS

The hypothesized model is described in Figure 5.5. An SEM analysis was performed on hypothetical data from 408 clients of an adoption agency (one mean score for each adoptive couple for each concept). The analysis was performed using AMOS version 18. The assumptions of no missing data, multivariate normality, linearity, and absence of multivariate outliers, and perfect multicollinearity were evaluated. Each of the aforementioned assumptions seems tenable.

The hypothesized model was tested with maximum likelihood estimation. The hypothesized model appears to be a good fit to the data. P(CMIN) equals .261, and the present model is not rejected. CMIN/DF equals 1.30. By this criterion the current model is not rejected. GFI equals .979. By this criterion the present model is not rejected. CFI equals .979. By this criterion the present model is not rejected. RMSEA equal to .00 − .05 indicates close fit. As PCLOSE equals .394, we do not reject the null hypothesis and conclude that RMSEA is no greater than .05. By this criterion the present model is not rejected. SRMR equals .0513. Therefore, the model is a fair to good fit. SRMR $\leq$ .08 suggests good fit. Additionally, the matrix of correlation was inspected. No unusual coefficients were observed. No modification index was greater than zero, and, consequently, no model modifications were suggested. No modification index was greater than zero, and, consequently, no model modifications were suggested.

In addition to considering overall model fit, it is important to consider the size of the estimated parameters, which are analogous to regression coefficients. As with regression, a model that fits the data quite well but has few practically significant parameters would be meaningless. At a minimum, the researcher should inspect model estimates to determine if proposed parameters were of practically significant magnitudes and in the expected directions.

For the final model (see Figure 5.16), correlation coefficients for exogenous variables, and standardized regression coefficients for endogenous variables were calculated. Of note for the exogenous variables, *Self Efficacy* and *Relationship satisfaction* were moderately correlated ($r = .47$). Of note for the endogenous variables, the standardized coefficients between *Relationship Satisfaction* and *Service Utilization* was $-.52$, *Self Efficacy* and *Service Utilization* was $-.15$, *Relationship Satisfaction* and *Satisfaction with Parenting* was $.16$, *Service Utilization* and *Satisfaction with Parenting* was $.19$, and *Satisfaction with Parenting* and *Satisfaction with Adoption Agency* was $.28$. That is, the best predictor of *Satisfaction with Parenting* is *Service Utilization*, and the best predictor of *Service Utilization* is *Sat with Parenting*.

ADDITIONAL EXAMPLES OF SEM FROM THE APPLIED RESEARCH LITERATURE

Benda, B. B. (2003). Test of a structural equation model of comorbidity among homeless and domiciled military veterans. *Journal of Social Service Research*, *29*(1), 1–35.

This exploratory study of 600 Vietnam era military veterans was designed to test a hypothesized theoretical model of comorbidity (substance abuse and depression) among domiciled and homeless persons. The model tested represented an integration of stress process and social support models found in the literatures on substance abuse and on depression. Caregiver attachment and early abuse also were used to elaborate on the integrated stress-social support theoretical model. Using **structural equation modeling**, all relationships in the hypothesized model were supported by data from domiciled veterans, except attachments to caregivers were not related to peer associations with substance users, and these associations were not related to depression. In contrast, all 24 relationships hypothesized in the model tested were supported among homeless veterans. Conceptual and practice implications of the findings were discussed.

Fandrem, H., Strohmeier, D., & Roland, E. (2009). Bullying and victimization among native and immigrant adolescents in Norway: The role of proactive and reactive aggressiveness. *The Journal of Early Adolescence, 29*(6), 898–923.

This study compares levels of bullying others, victimization, and aggressiveness in native Norwegian and immigrant adolescents living in Norway and shows how bullying is related to proactive and reactive aggressiveness. The sample consists of 2,938 native Norwegians and 189 immigrant adolescents in school grades 8, 9, and 10. Data were collected via self-assessments. **SEMs** were conducted separately for girls and boys in both groups. The levels of victimization, reactive and proactive aggressiveness were the same for both native Norwegians and immigrant adolescents but there was a significant difference in the levels of bullying others. Compared with the native Norwegians, immigrant adolescents were found to be at higher risk of bullying others. Structural models revealed significantly stronger relations between affiliation-related proactive aggressiveness and bullying others in immigrant boys compared with the other groups. This indicates that the wish for affiliation is an important mechanism of bullying others in immigrant boys. The authors also suggest further research and the practical importance of the findings for prevention of targeting immigrant adolescents.

Jung, J. S. (2009). The relationships of flexible work schedules, workplace support, supervisory support, work-life balance, and the well-being of working parents. *Journal of Social Service Research, 35*(2), 93–104.

This study, using a secondary dataset from the 2002 National Study of the Changing Workforce, examines how working parents cope with work demands and family responsibilities. The design is a study on the relationships of flexible work schedule, workplace support, supervisory support, and work-life balance on the well-being of working parents employing the **SEM**. In this study, *employee well-being* is an endogenous latent construct. *Work-schedule flexibility, workplace support, supervisory support,* and *work-life balance* are latent exogenous constructs. This information will assist social workers in developing more effective intervention efforts in the workplace, with the ultimate goal of increasing the quality of life.

McGowan, B. G., Auerbach, C., & Strolin-Goltzman, J. S. (2009) Turnover in the child welfare workforce: A different perspective. *Journal of Social Service Research, 35*(3), 228–235.

Child welfare agencies across the country are experiencing a workforce crisis involving high staff turnover rates. The purpose of this study was to determine which of the organizational, personal, and supervisory variables identified in prior research on this topic are most associated with intent to leave among employees in urban and rural child welfare settings. Four-hundred-and-forty-seven employees in 13 child welfare agencies participated in a survey addressing organizational, personal, and supervisory factors related to turnover. ANOVA, logistical regression, and *SEM* were used in the data analysis. The organizational and supervisory variables identified as significant in the logistic regression, as in earlier research, were not significant when the data were subjected to structural equation modeling. Instead, findings suggest that career satisfaction and satisfaction with paperwork are the key determinants of workers' intention to stay.

Moon, S. S. (2009). Ecological influences on school achievement in a diverse youth sample: the mediating role of substance use. *Journal of Human Behavior in the Social Environment, 19*(5), 572–591.

This study's purpose was to examine the extent to which closeness to family, peers, and school was associated with substance use and school achievement, based on the integrative model of ecological theory, social attachment theory, and social learning theory. A secondary data analysis was conducted on the first wave of the National Longitudinal Study of Adolescent Health. The final sample yielded 3,147 boys and 3,356 girls. A **SEM** was employed to test a hypothesized model. School closeness was found to be a primary ecological factor that significantly influenced school achievement while substance use emerged as a critical mediator of this relationship. Family closeness was negatively associated with school achievement. Also, substance use did not have a mediating function in the relationship between peer closeness and school achievement. No gender difference was found, except the relationship between family closeness and school achievement, in that family closeness had a significant, direct effect on school achievement among only boys but not girls.

Owens, T. J. (2009). Depressed mood and drinking occasions across high school: comparing the reciprocal causal structures of a panel of boys and girls. *Journal of Adolescence, 32* (4), 763–780.

Does adolescent depressed mood portend increased or decreased drinking? Is frequent drinking positively or negatively associated with emotional well-being? Do the dynamic relations between depression and drinking differ by gender? Using block-recursive **SEMs**, we explore the reciprocal short-term effects (within time, t) and the cross-lagged medium-term effects ($t + 1$ year) and long-term effects ($t + 2$ years) of depressed mood and monthly drinking occasions. Data come from the high school waves of the Youth Development

Study, a randomly selected panel of 1015 ninth graders followed to 12th grade. The author found that for both genders, depressed mood consistently decreased short-term drinking in each grade measured. However, depression increased drinking for both genders in the medium-term but only for girls in the long-term. In the other direction, drinking tended to increase depression in the short-term only among ninth-grade boys and 12th-grade girls. Observed trends and differences in the magnitude of the reciprocal effects vary by gender, with drinking being especially deleterious to emotional well-being for boys early in high school (10th grade) but for girls on the cusp of the post-high school world (12th grade).

6

Choosing among Procedures for the Analysis of Multiple Dependent Variables

---

Chapters 2, 3, and 4 have discussed the following traditional or first-generation statistical tools: MANOVA, MANCOVA, and MMR, respectively. Chapter 5 discussed SEM, which is a more recent or second generation addition to the statistical toolbox of social workers and other applied researchers. In one sense, the aforementioned multivariate procedures are variations on a single theme: the general linear model. Nonetheless, there are differences among these procedures that make each more appropriate given the (1) purpose of the analysis; (2) sample size; (3) tenability of the GLM's assumptions; and (4) type of DV system being modeled.

This chapter (1) summarizes similarities and differences between MANOVA, MANCOVA, MMR, and SEM, and (2) offers suggestions to guide their differential use. Figure 6.1 compares these four multivariate procedures on selected criteria. Remaining discussion compares and contrasts MANOVA and MANCOVA versus MMR, MANOVA and MANCOVA versus SEM, and MMR versus SEM.

Criterion	MANOVA	MANCOVA	MMR	SEM
Research Objective	Focuses on mean differences	Focuses on mean differences while controlling for other variable that may affect these differences	Test models that focuses on relationships between a set of IVs and a set of DVs	Test models that Focus on individual paths between exogenous and endogenous variables
Important Assumptions	GLM	GLM	GLM	GLM
Variables Modeled	Multiple Independent (IV) Multiple Dependent (DV)	Multiple Independent (IV) Multiple Dependent (DV)	Multiple Independent (IV) Multiple Dependent (DV)	Multiple Exogenous Multiple Endogenous
Level of Measurement	IVs = Nominal DVs = Interval/Ratio	IVs = Nominal DVs = Interval/Ratio Covariates = Interval/Ratio	IVs = All DVs = Interval/Ratio	Exogenous = All Endogenous = All
Minimum Sample Size	Smaller	Smaller	Smaller	Larger
Criteria to Evaluate the Utility of the Model Tested	Eta Squared	Eta Squared	Multiple R-squared	Various, including model chi square, CFI, RMSEA
Modeling Capabilities	Path of all IVs to a linear combination of all DVs	Path of all IVs to a linear combination of all DVs, while controlling for other variable that may affect these differences	Path of all IVs to a of each DV	Paths to observed and latent endogenous variables
Type of Variance Modeled	Common	Common	Common	Common Specific Error
Moderating Relationships Modeled?	Yes	Yes	Yes	Yes
Mediating Relationships Modeled?	No	No	Yes	Yes
Causation Established?	No	No	No	No

Figure 6.1 Comparison of Four Multivariate Procedures on Selected Criteria.

MANOVA AND MANCOVA VERSUS MMR

As stated above, MANOVA shares the same underlying mathematical model with linear regression: the GLM. Neither MANOVA/MANCOVA nor MMR make assumptions about the distributions of the IVs. Consequently, the individual IVs do not need to be normally distributed in order to perform MANOVA/MANCOVA or a MMR. Both MANOVA/MANCOVA and MMR assume that a model's residuals (ε) are normally distributed random variables with the same variance (i.e., homoscedastic). Heteroscedasticity can be problematic in all three of the aforementioned procedures. With MMR, strong heteroscedasticity may cause the variance around the estimated slope and intercept to be underestimated (Miller, 1986), potentially leading to overestimates of statistical significance. In MANOVA/MANCOVA, heteroscedasticity alters the assumptions underlying the F-test and may cause the p-values to be over- or underestimated (Miller, 1986).

Although they have the same underlying mathematical framework, MANOVA/MANCOVA and MMR are different in several important ways. MANOVA and MANCOVA seek to answer different questions, and, consequently, test different hypotheses compared with MMR. MMR quantitatively describes the relationship between a DV and one or more IVs. MANOVA and MANCOVA describe whether a set of DVs differs among three or more levels of an IV. Regression has been applied most often to data obtained from correlational or nonexperimental research to describe and predict DV values on the basis of a model constructed from the relationship between IVs and DVs. In contrast, MANOVA and MANCOVA have been applied most frequently to experimental data. Imprecise or inaccurate estimates of the independent variables are a particular concern for regression, which explicitly assumes that all predictors are measured without error, or at least with much less error than the response variable Y. Violation of this assumption leads to ***errors in variables*** (EIV) and biased parameter estimates. For example, in simple linear regression, EIV bias regression slopes toward zero (Sokal & Rohlf, 1995), potentially altering biological conclusions and complicating the use of regression models in further research. Most important, the general linear model assumes that the relationship between Y and X can be described using a linear equation, so that regression is inappropriate when the relationship cannot be made linear in the parameters (e.g., through transformations or polynomial terms).

Cottingham, Lennon, and Brown (2005) argue that because linear regression, and, consequently, MMR, requires fewer parameters to model a relationship, it is generally a more powerful statistical approach than MANOVA/MANCOVA. As discussed above, power is defined as the probability of detecting an effect when that effect is present (i.e., the probability of rejecting the null hypothesis when the null hypothesis is false). Assuming constant values for sample size, alpha, and beta, the fewer the parameters (i.e., IVs in the model), the greater the statistical power of an analysis.

For simplicity, the following example focuses on ANOVA and regression. This example can be extended to include MANOVA/MANCOVA and MMR. Suppose that a researcher designs a study to quantify the effect of an intervention on the level of depression of a study's participants by using five levels of that intervention (i.e., 1 session, 5 sessions, 10 sessions, 15 sessions, and 20 sessions). A simple linear regression to account for the effect of the intervention would have two parameters: intercept and slope. On the other hand, a one-way ANOVA model for the same study would require five parameters, each specifying the mean for an intervention level. This difference in the number of parameters grows as the number of interventions increases. There is one important exception. When there are only two levels per factor (i.e., per IV), the power of ANOVA is always equivalent to the power of regression because both have the same number of parameters. Accordingly, a one-way ANOVA with two levels of the IV has the same power as a simple (i.e., bivariate) linear regression, and a two-way ANOVA with two levels per factor has the same power as a multiple regression model with main effects and an interaction term.

MANOVA AND MANCOVA VERSUS SEM

MANOVA often is recommended when a set of DVs constitutes a variable system, also referred to as a factor or latent variable (e.g., Huberty & Morris, 1989). For example, a researcher may measure a factor such as social interaction with three indicators: time spent with friends, time spent with family, and time spent with coworkers. An alternative way to conduct an analysis when a set of DVs constitutes a variable system is to compare group means with SEM (Kline, 2011). Cole, Maxwell, Arvey, and

Salas (1993), for example, have argued that the choice between MANOVA and SEM should be guided by the question under investigation *and* by the type of DV system being modeled.

More specifically, both SEM and MANOVA may be used to test models that link latent variables (or factors) and the empirical indicators of those latent variables. But, the type of link between latent variables and empirical indicators differs for the two approaches. In SEM models, the direction is from the latent variables to the indicators; in contrast, in MANOVA models, the direction is from the empirical indicators to the latent variables. Following Bollen (1989) and Bollen and Lennox (1991), in SEM, empirical indicators of a latent variable are referred to as *effect indicators,* and in MANOVA, empirical indicators of a latent variable are referred to *cause indicators*. It is crucial in deciding which analysis to conduct to distinguish between DVs (i.e., exogenous variables) that are affected by latent variables and DVs (i.e., exogenous variables) that cause latent variables.

While it may seem reasonable to expect indicators of the same latent variable to be positively related with each other, indicators of a latent variable may not be related to each other. More specifically, empirical indicators should be related to each other if they are the effects of a latent variable. For example, to measure self-esteem, a person may be asked to indicate whether he or she agrees or disagrees with the statements: (1) I believe that I am a good person; and (2) I am happy with who I am. A person with high self-esteem should agree with both statements; in contrast, a person with low self-esteem would probably disagree with both statements. Because each indicator depends on or is caused by self-esteem, both of the aforementioned indicators should be positively correlated with each other. That is, indicators that depend on the same variable should be associated with one another if they are valid measures (Rubin, 2010). In contrast, when indicators are the cause rather than the effect of a latent variable, these indicators may correlate positively, correlated negatively, or be uncorrelated (Rubin, 2010). For example, gender and race could be used as indicators of the variable "exposure to discrimination." Being nonwhite or female increases the likelihood of experiencing discrimination, so both are good indicators of the variable. But, race and gender of individuals would not be expected to be strongly associated.

In summary, in SEM, empirical indicators (i.e., measured variables) are hypothesized to be linear combinations of latent variables plus error;

that is, arrows are directed from latent to empirical indicators. In SEM, measured variables are *effect indicators* within a latent variable system in that the indicators are affected by the factors (Bollen &Lennox, 1991; MacCallum & Browne, 1993). In MANOVA and MANCOVA, researchers may hypothesize that a latent variable (e.g., program effectiveness) is a linear combination of measured variables (e.g., increased client self-efficacy, congruence between client expectations and program performance, and client reports of program responsiveness in terms of respectful and timely worker behaviors). Under these conditions, measured variables may be best conceptualized as *cause indicators* (Bollen & Lennox. 1991; MacCallum & Browne, 1993).

MMR VERSUS SEM

First generation multivariate methods, such as MMR, are appropriate for evaluating constructs and relationships between constructs. As described earlier, MMR is an extension of OLS regression. MMR estimates the same regression coefficients and standard errors that would be obtained by using separate OLS regressions equations for each DV. However, MMR also may be used to test an omnibus null hypothesis and composite hypotheses for a model. The *omnibus null hypothesis* is that all regression coefficients equal zero across all DVs. Tests of individual IVs or subsets of IVs across DVs are termed *composite hypotheses* (Cramer & Nicewander, 1979).

In MMR, the variables used to specify and test a model are assumed to be measured without error, error terms (or residuals) are not intercorrelated, and variables in the model are unidirectional (does not incorporate feedback loops among variables). These assumptions are rarely, if ever, met in nonexperimental social research. In MMR, the simultaneous evaluation of model construct relationships is not possible; evaluation has to be performed in sequential steps. In contrast, SEM allows the simultaneous analysis of all the variables in a model. In addition, SEM is capable of testing relationships between latent and observed variables.

Given the advantages of SEM over OLS regression, when would one ever want to use MMR? Jaccard and Wan (1996) state that regression may be preferred to SEM when there are substantial departures from the SEM assumptions of multivariate normality of the indicators and small sample sizes, SEM requires relatively large samples. Moreover, SEM focuses on testing overall models, whereas significance tests are of single effects. While many of the measures used in SEM can be assessed for significance, significance testing is less important in SEM than in other multivariate techniques. In other techniques, significance testing is usually conducted to establish that we can be confident that a finding is different from the null hypothesis, or, more broadly, that an effect can be viewed as "real." In SEM the purpose is usually to determine if one model conforms to the data better than an alternative model. It is acknowledged that establishing this does not confirm "reality" as there is always the possibility that an unexamined model may conform to the data even better. More broadly, in SEM the focus is on the strength of conformity of the model with the data, which is a question of association, not significance.

In conclusion, researchers prefer to make causal inferences from randomized experiments. However, often, experiments are impractical or unethical. Consequently, a significant proportion of our social science knowledge is derived from nonexperimental (i.e. observational) studies. In nonexperimental investigations, judgment is required to assess the probable impact of measured and unmeasured bias such as confounding variables, and mathematical equations (i.e., statistical models) are frequently used to adjust for confounding and other sources of bias. The challenges of making causal statements based on nonexperimental data are well known. However, it is not that causal statements based on nonexperimental data are impossible, but that they usually are difficult make.

The perspective taken here is that statistical models (i.e., multivariate procedures) are an important design feature in observation or nonexperimental studies. For example, in a SEM context, even a correctly specified and tested theoretical model (e.g., one that includes all necessary variables and paths) can fit poorly and yield highly biased estimates if the study is poorly designed. Multivariate procedures allow social workers and other human services researchers to analyze multidimensional social problems and interventions in ways that minimize oversimplification. Multivariate procedures allow researchers to use multiple operationalism, which refers to the use of two or more measures to represent a concept. The use of multiple measures of a single concept provides a better chance of fully representing that concept. A second reason to conduct a multivariate analysis is to control type I error. However, there are costs associated with the use of multivariate procedures. The use of too many DVs may reduced power or result in spurious findings due to chance.

The proper selection of an analytical strategy is a crucial part of the research study. Which strategy is most appropriate depends on the: (1) purpose of the analysis; (2) sample size; (3) tenability of assumptions; and (4) type of DV system being modeled. These three issues have been discussed for MANOVA, MANCOVA, MMR, and SEM, which were presented as alternative statistical procedures for analyzing models with more than one DV.

References

Anderson, T. W. (1999). Asymptotic theory for canonical correlation analysis. *Journal of Multivariate Analysis, 70,* 1–29.

Anderson, T. W. (2002). Canonical correlation analysis and reduced rank regression in autoregressive models. *Annals of Statistics, 30,* 1134–1154.

Babyak, M. A. (2004). What you see may not be what you get: A brief, nontechnical introduction to overfitting in regression-type models. *Psychosomatic Medicine, 66,* 411–421.

Bagozzi, R. P., & Edwards, J. R. (1998). A general approach for representing constructs in organizational research. *Organizational Research Methods, 1,* 45–87.

Bagozzi, R. P., Fornell, C., & Larcker, D. F. (1981). Canonical correlation analysis as a special case of a structural relations model. *Multivariate Behavioral Research, 16,* 437–454.

Bagozzi, R. P., Yi, Y, & Singh, S. (1991). On the use of structural equation models in experimental designs: Two extensions. *International Journal of Research in Marketing, 8,* 125–140.

Baggaley, A. R. (1981). Multivariate analysis: An introduction for consumers of behavioral research. *Evaluation Review, 5,* 123–131.

Barcikowski, R., & J. P. Stevens . (1975). A Monte Carlo study of the stability of canonical correlations, canonical weights, and canonical variate-variable correlations. *Multivariate Behavioral Research, 10,* 353–364.

Bandalos, D. L. (2002). The effects of item parceling on goodness-of-fit and parameter estimate bias in structural equation modeling. *Structural Equation Modeling, 9,* 78–102.

Bartlett, M.S. (1938). Further aspects of the theory of multiple correlation. *Proceedings of the Cambridge Philosophical Society, 34,* 33–40.

Bartlett, M. S. (1948). Internal and external factor analysis. *British Journal of Psychology*, *1*, 73–81.

Bartlett, M. S. (1951). The effect of standardization on a approximation in factor analysis. *Biometrika*, *38*, 337–344.

Barrera, M., & Garrison-Jones, C. (1992). Family and peer social support as specific correlates of adolescent depressive symptoms. *Journal of Abnormal Child Psychology*, *20*(1), 1–16.

Bentler, P. M. (1980). Multivariate analysis with latent variables: Causal modeling. *Annual Review of Psychology*, *31*, 419–456.

Bentler, P. M. (1990). Comparative fit indexes in structural models. *Psychological Bulletin*, *107*(2), 238–246.

Bentler, P. M., & Weeks, D. G. (1980). Linear structural equations with latent variables. *Psychometrika*, *45*, 289–308.

Berg, A. J., Ingersoll, G. M., & Terry, R. L. (1985). *Psychological Reports*, *56*(1), 115–122.

Berk, R. A. (2003). *Regression analysis: A constructive critique*. Thousand Oaks, CA: Sage Publications.

Bickel, Robert. (2007). *Multilevel analysis for applied research: It's just regression!* New York: Guilford Press.

Bollen, K. A. (1989). *Structural equations with latent variables*. New York: Wiley.

Bollen, K., & Lennox, R. (1991). Conventional wisdom on measurement: A structural equation perspective. *Psychological Bulletin*, *110*, 305–314.

Boomsma, A. (2000). Reporting analyses of covariance structures. *Structural Equation Modeling*, *7*, 461–483.

Boring, E. G. (1953). The role of theory in experimental psychology. *The American Journal of Psychology*, *69*(2), 169–184.

Bray, J. H., & Maxwell, S. E. (1985). *Multivariate analysis of variance*. Newbury Park, CA: Sage Publications.

Breivik, E., & Olsson, U. H. (2001). Adding variables to improve fit: The effect of model size on fit assessment in LISREL. In R. Cudeck, S. Du Toit, & D. Sörbom (Eds.), *Structural equation modeling: Present and future* (pp. 169–194). Lincolnwood, IL: Scientific Software International.

Bryk, A., S. W., Raudenbush, Seltzer, M., & Congdon, R. T. (1988). *An introduction to HLM*. Chicago: University of Chicago.

Bring, J. (1994). How to standardize regression coefficients. *The American Statistician*, *48*(3), 209–213.

Brown, T. A. (2006), *Confirmatory factor analysis for applied research*. New York: Guilford Press.

Brown, S. R., & Melamed, L. M. (1990). *Experimental design and analysis*. Thousand Oaks, CA: Sage Publications.

Bruning, J. M., & Kintz, B. L. (1987). *Computational handbook of statistics.* Glenview, Illinois: Scott, Foresman and Company.

Campbell, D. T., & Fiske, D. W. (1959). Convergent and discriminant validation by the multitrait-multimethod matrix. *Psychological Bulletin, 56*(2), 81–103.

Cargill, B. R., Emmons, K. M., Kahler, C.W., & Brown, R. A. (2001). Relationship among alcohol use, depression, smoking behavior, and motivation to quit smoking with hospitalized smokers. *Psychology of Addictive Behaviors, 15*(3), 272–275.

Carroll, J. D., & Green, P. E. (1997). *Mathematical tools for applied multivariate analysis.* New York: Academic Press.

Cattell, R. J. (1956). Validation and intensification of the sixteen personality factor questionnaire. *Journal of Clinical Psychology, 12*, 205–214.

Coanders, G., Satorra, A., & Saris, W. E. (1997). Alternative approaches to structural equation modeling of ordinal data: A Monte Carlo study. *Structural Equation Modeling, 4*, 261–282.

Cohen, J., Cohen, P., West, S. G., & Aiken, L. S. (2003). Applied multiple regression/correlation analysis for the behavioral sciences. Mahwah, NJ: Erlbaum.

Cohen, J., & Cohen, P. (1983). Applied multiple regression/correlation analysis for the behavioral sciences. Hillsdale, NJ: Erlbaum.

Cohen, J. (1988). *Statistical power analysis for the behavioral sciences* (2nd ed.). Hillsdale, NJ: Erlbaum.

Cohen, J. (1994). The earth is round (p < .05). *American Psychologist, 49*(12), 997–102.

Cleroux, R., & Lazraq, A. (2002). Testing the signicance of the successive components in redundancy analysis. *Psychometrika, 67*, 411–419.

Cliff, N. (1983). Some cautions concerning the application of causal modeling methods. *Multivariate Behavioral Research, 18*, 81–105.

Cliff, N., & Krus, J. D. (1976). Interpretation of canonical analysis: Rotated vs. unrotated solutions. *Psychometrika, 41*, 35–42.

Cocozzelli, C., & Constable, R T. (1985). An empirical analysis of the relation between theory and practice in clinical social work. *Journal of Social Service Research, 9*(1), 47–64.

Cohen, J., & Cohen, P. (1983). *Applied multiple regression/correlation analysis for the behavioral sciences.* New Hersey: Lawrence Erlbaum Associates.

Cohen, J. (1988). *Statistical power analysis for the behavioral sciences* (2nd ed.). Hillsdale, NJ: Erlbaum.

Cole, D. A., Maxwell, S. E., Arvey, R., & Salas, E. (1993). Multivariate group comparisons of variable systems. *Psychological Bulletin, 114*(1), 174–184.

Cole, D. A., & Maxwell, S. E. (1985). Multitrait-multimethod comparisons across populations: A confirmatory factor analytic approach. *Multivariate Behavioral Research, 20*, 389–417.

Cook, J. A., Razzano, L., & Cappelleri, J. C. (1996). *Canonical correlation analysis of residential and vocational outcomes following psychiatric rehabilitation. Evaluation and Program Planning, 19*(4), 351–363.

Cooley, W. W., & Lohnes, P. R. (1971). *Evaluation research in education.* New York: Irvington.

Cottingham, K. L., Lennon, J. T., & Brown, B. L. (2005). Knowing when to draw the line: designing more informative ecological experiments. *Frontiers in Ecology and the Environment, 3*(3), 145–152.

Cox, D. R., & Small N. J. H. (1978). Testing multivariate normality. *Biometrika 65*(2), 263–272.

Cramer, E. M., & Nicewander, W. A. (1979). Some symmetric, invariant measures of multivariate association. *Psychometrika, 44*, 43–54.

Chrisman, N. R. (1998). Rethinking levels of measurement for cartography. *Cartography and Geographic Information Science, 25*(4), 231–242.

DeCarlo, L. T. (1997). On the meaning and use of kurtosis. *Psychological Methods, 2*(3), 292–307.

Draper, N. R. and Smith, H. (1998). *Applied Regression Analysis.* New York: John Wiley.

Draper, N. R., Guttman, I., & Lapczak, L. (1979). Actual rejection levels in a certain stepwise test. *Communications in Statistics, 8*, 99–105.

Duffy, M. E., Wood, R. Y., & Morris, S. (2001). The influence of demographics, functional status and comorbidity on the breast self-examination proficiency of older African-American women. *Journal of National Black Nurses Association, 12*(1), 1–9.

Edgeworth, F. Y. (1886). Progressive means. *Journal of the Royal Statistical Society, 49*, 469–475.

Enders, C. K. (2003). Performing multivariate group comparisons following a statistically significant MANOVA. *Measurement and Evaluation in Counseing and Development, 36*, 40–56.

Enders, C. K. (2010). Applied missing data analysis. New York: The Guilford Press.

Fan, X. (1997). Canonical correlation analysis and structural equation modeling: What do they have in common? *Structural Equation Modeling, 4*, 65–79.

Fan, X., Thompson, B., & Wang, L. (1999). The effects of sample size, estimation methods, and model specification on SEM fit indices. *Structural Equation Modeling, 6*, 56–83.

Finch, W. H. (2007). Performance of the Roy-Bargmann stepdown procedure as a follow up to a significant MANOVA. *Multiple Regression Viewpoints, 33*(1), 12–22.

Finn, J. D. (1974). *A general model for multivariate analysis.* New York: Holt.

Finn, J. (1978). *Multivariance: Univariate and multivariance analysis of variance, covariance and regression.* Chicago: National Education Resources.

Fisher, R. A. (1918). The correlation between relatives on the supposition of Mendelian inheritance. *Philosophical Transactions of the Royal Society of Edinburgh, 52,* 399–433.

Fisher, R. A. (1925). *Statistical methods for research workers.* Edinburgh, England: Oliver and Boyd.

Fisher, R. A. (1935). *The design of experiments.* Edinburgh, England: Oliver and Boyd.

Flores, L. Y., Navarro, R. L., & DeWitz, S. J. (2008). Mexican American high school students' postsecondary educational goals: Applying social cognitive career theory. *Journal of Career Assessment, 16*(4), 489–501.

Fox, J. (1991). *Regression diagnostics.* Thousand Oaks, CA: Sage Publications.

Friedman, J. H., & Rafsky, L. C. (1979). Multivariate generalizations of the Wald-Wolfowitz and Smirnov two sample tests. *Annals of Statistics, 7,* 697–717.

Galton, F. (1886). Regression toward mediocrity in hereditary stature. *Journal of the Anthropological Institute, 15,* 246–263.

Gelman, A. (2008). Scaling regression inputs by dividing by two standard deviations. *Statistics in Medicine, 27,* 2865–2873.

Gunst, R. F., & Mason, R. (1980). *Regression analysis and its applications: A data-oriented approach.* New York: Marcel Dekker, Inc.

Guo, B., Perron, B. E., & Gillespie, D. F. (2009). A systematic review of structural equation modelling in social work research. *British Journal of Social Work, 39,* 1556–1574.

Hair, J. F., Anderson, R. E., Tatham, R. L., & Black, W. C. (1998). *Multivariate data analysis.* New York: Prentice Hall.

Hall, R. J., Snell, A. F., Foust, M. S. (1999). Item parceling strategies in SEM: Investigating the subtle effects of unmodeled secondary constructs. *Organizational Research Methods, 2,* 757–765.

Hayduk, L. A. (1996). *LISREL issues, debates and strategies.* Baltimore: Johns Hopkins University Press.

Heitjan, D. (1997). Annotation: What can be done about missing data? Approaches to imputation. *American Journal of Public Health, 87,* 548–550.

Henderson, M. J., Saules, K. K., & Galen, L. W. (2004). The predictive validity of the University of Rhode Island Change Assessment Questionnaire in a heroin-addicted polysubstance abuse sample. *Psychology of Addictive Behaviors, 18*(2), 106–112.

Hocking, R. R. (1976). The analysis and selection of variables in linear regression. *Biometrics, 32*(1), 1–49.

Hotelling, H. (1931). The generalization of Student's ratio. *Annals of Mathematical Statistics, 2*(3), 360–378.

Hotelling, H. (1935). The most predictable criterion. *Journal of Educational Psychology, 26*, 139–142.

Hotelling, H. (1936). Relations between two sets of variates. *Biometrika, 28*, 321–377.

Hsu, J. (1996). *Multiple comparisons: Theory and methods.* New York, NY: Chapman & Hall.

Howell, D. C. (2009). *Statistical methods for psychology.* Belmont, CA: Wadsworth.

Hox, J. J. (2002). *Multilevel analysis: Techniques and applications.* Mahwah, NJ: Lawrence Erlbaum.

Hoyle, R. H., & Panter, A. T. (1995). Writing about structural equation models. In R. H. Hoyle (Ed.), *Structural equation modeling* (pp. 158–176). Thousand Oaks, CA: Sage.

Hu, L.-T., & Bentler, P. (1995). Evaluating model fit. In R. H. Hoyle (Ed.), *Structural Equation Modeling. Concepts, Issues, and Applications* (pp. 76–99). London: Sage.

Hu, Li-tze, & Bentler, P. M. (1999). Cutoff criteria for fit indexes in covariance structure analysis: conventional criteria versus new alternatives. *Structural Equation Modeling, 6*(1), 1–55.

Huberty, C.J. (1994). *Applied discriminant Analysis.* New York: John Wiley and Sons, Inc.

Hubety, C. J. (1994a). Why multivariate analyses? *Educational and Psychological Measurement, 64*, 620–627.

Huberty, C., & Morris, J. D. (1989). Multivariate analysis versus multiple univariate analyses. *Psychological Bulletin, 105*(2), 302–308.

Huberty, C. J., & Olejnik, S. (2006). *Applied MANOVA and discriminant Analysis.* New Jersey: John Wiley & Sons, Inc.

Huberty, S. J., & Smith, J. D. (1982). The study of effects in MANOVA. *Multivariate Behavioral Research, 17*, 417–432.

Huitema, B. E. (1980). *The analysis of covariance and alternatives.* New York: Wiley.

Hunter. J. E., & Gerbing, D. W. (1982). Unidimensional measurement, second-order factor analysis, and causal models. In B. M. Staw & L. L. Cummings (Eds.), *Research in organizational behavior* (pp. 267–299). Greenwich, CT: JAI Press.

Jaccard J., & Guilamo-Ramos V. (2002). Analysis of variance frameworks in clinical child and adolescent psychology: issues and recommendations. *Journal of Clinical Child and Adolescent Psychology, 31*(1):130–46.

Jaccard, J., & Wan, C. K. (1996). *LISREL analyses of interaction effects in multiple regression.* Newbury Park, CA: Sage.

Jagannathan, R., & Camasso, M. J. (1996). Risk assessment in child protective services: A canonical analysis of the case management function. *Child Abuse and Neglect, 20*(7), 599–612.

Jöreskog, K. G. (1971). Statistical analysis of sets of congeneric tests. *Psychometrika, 36*, 109–133.

Jöreskog, K. G. (2004). *On chi-squares for the independence model and fit measures in LISREL.* Retrieved from December 27, 2010 from http://www.ssicentral.com/lisrel/techdocs/ftb.pdf

Joreskog, K. G., & Sorbom, D. (1991). *LISREL 7: A guide to the program and applications.* Chicago: SPSS.

Kane, M. N., Hamlin, E. R. & Hawkins, W. E. (2002). Which clinical methods are associated with better preparing students and practitioners for managed care, state licensing, and other important areas? *Professional Development: The International Journal of Continuing Social Work Education, 5*(2), 15–27.

Kaplan, D. (2000). *Structural equation modeling: Foundations and extensions.* Thousand Oaks, CA: Sage.

Keith, T. Z. (2006). *Multiple regression and beyond.* New York: Pearson.

Kenny, D. A. & Judd, C. M. (1986). Consequences of violating the independence assumption in analysis of variance. *Psychological Bulletin, 99*, 422–432.

Kerlinger, F. N., & Pedhazur, E. J. (1973) *Multiple regression in behavioral research.* New York: Holt, Rinehart & Winston.

Kim, J. O. (1981). Standardization in causal analysis. *Sociological Methods and Research, 10*, 187–210.

Kishton, J. M., & Widaman, K. F. (1994). Unidimensional versus domain representative parceling of questionnaire items: An empirical example. *Educational and Psychological Measurement, 54*, 757–765.

Kline, R. B. (2011). *Principles and practice of structural equation modeling.* New York: Guilford Press.

Klockars, A. J. and Sax, G. (1986). *Multiple comparisons.* Thousand Oaks, CA: Sage Publications.

Klockars, A. J., Hancock, G. R., & McAweeney, M. J. (1995). Power of unweighted and weighted versions of simultaneous and sequential multiple-comparison procedures. *Psychological Bulletin, 118*, 300–307.

Knapp, T. R. (1978). Canonical correlation analysis: A general parametric significance testing system. *Psychological Bulletin, 85*(2), 410–416.

Kowlowsky, M., & Caspy, T. (1991). Stepdown analysis of variance: A refinement. *Journal of Organizational Behavior, 12*, 555–559.

Kromrey, J. D., & La Rocca, M. A. (1995). Power and Type I error rates of new pairwise multiple comparison procedures under heterogeneous variances. *Journal of Experimental Education, 63*, 343–362.

Larsen, J. J., & Juhasz, A. M. (1985). The effects of knowledge of child development and social-emotional maturity on adolescent attitudes toward parenting. *Adolescence*, *20*(80), 823–39.

Leary, M. R., & Altmaier, E. M. (1980). Type I error in counseling research: A plea for multivariate analyses. *Journal of Counseling Psychology*, *27*, 611–615.

Loehlin, J. C. (2004). *Latent variable models: An introduction to factor, path, and structural equation analysis.* Mahwah, NJ: Erlbaum.

Lorenz, Frederick O. (1987). Teaching about influence in simple regression. *Teaching Sociology*, *15*(2), 173–177.

Little, R. J., & Rubin, D. B. (2002). *Statistical analysis with missing data.* New York: John Wiley & Sons.

Little, T. D., Cunningham, W. A., Shahar, G., & Widaman, K. F. (2002). To parcel or not to parcel: Exploring the question, weighing the merits. *Structural Equation Modeling*, *9*, 151–173.

Loehlin, J. C. (2004). *Latent variable models: An introduction to factor, path, and structural equation analysis.* Mahwah, NJ: Erlbaum.

Looney, S. W. (1995). How to use tests for univariate normality to assess multivariate normality. *The American Statistician*, *49*(1), 64–70.

Lorenz, F. O. (1987). Teaching influence in simple regression. *Teaching Sociology*, *15*(2), 173–177.

Lynch, S. M., & Graham-Bermann, S. A. (2004). Exploring the relationship between positive work experiences and women's sense of self in the context of partner abuse. *Psychology of Women Quarterly*, *28*(2), 159–167.

MacCallum, R. C., & Austin, J. T. (2000). Applications of structural equation modeling in psychological research. *Annual Review of Psychology*, *51*, 201–226.

MacCallum, R. C., & Browne, M. W. (1993). The use of causal indicators in covariance structure models: Some practical issues. *Psychological Bulletin*, *114*, 533–541.

MacCallum, R. C., Wegener, D. T., Uchino, B. N., & Fabrigar L. R. (1993). The problem of equivalent models in applications of covariance structure analysis. *Psychological Bulletin*, *114*, 185–99.

MacCallum, R. C., Widaman, K. F., Zhang, S., & Hong, S. (1999). Sample size in factor analysis. *Psychological Methods*, *4*, 84–99.

Mallows, C. L. (1973). Some comments on CP. *Technometrics*, *15*(4), 661–675.

Marsh, H. W., & O'Neill, R. (1984). Self Description Questionnaire III: The construct validity of multidimensional self-concept ratings by late adolescents. *Journal of Educational Measurement*, *21*, 153–174.

Martens, M. P. (2005). Future directions of structural equation modeling in counseling psychology. *The Counseling Psychologist*, *33*(3), 375–382.

Menard, S. (1995). *Applied logistic regression analysis.* Thousand Oaks, CA: Sage Publications.

Mardia, K. V. (1970). Measures of multivariate skewness and kurtosis with applications. *Biomnetrika, 57*, 519–530.

Maxwell, S. E., & Delaney, H. D. (2003). *Designing experiments and analyzing data: A model comparison perspective.* New York: Routledge.

McDonald, R. P. (1982). A note on the investigation of local and global identifiability. *Psychometrika, 47*(1), 101–103.

McDonald, R. P., & Ho, M. H. R. (2002). Principles and practice in reporting structural equation analyses. *Psychological Methods, 7*, 64–82.

Miller, R. G. (1986). *Beyond ANOVA: basics of applied statistics.* New York, NY: Wiley Press.

Miville, M. L., & Constantine, M. G. (2006). Sociocultural predictors of psychological help-seeking attitudes and behavior among Mexican American college students. *Cultural Diversity and Ethnic Minority Psychology, 12*(3), 420–432.

Miville, M. L., Darlington, P., Whitlock, B., & Mulligan, T. (2005). Integrating identities: The relationships of racial, gender, and ego identities among white college students. *Journal of College Student Development, 46*(2), 157–175.

Mudholkar, G. S., & Subbaiah, P. (1980). A review of step-down procedures for Multivariate Analysis of Variance. In R. P. Gupta, (Ed.), *Multivariate Statistical Analysis* (pp.161–178*)*. Amsterdam: North-Holland.

Mudholkar, G. S., & Subbaiah, P. (1988). On a fisherian detour of the step-down procedure for MANOVA. *Communications in Statistics: Theory and Methods, 17*, 599–611

Mulaik, S. A. (2009) *Linear causal modeling with structural equations.* New York: Chapman & Hall.

Muller, K. E. (1981). Relationships between redundancy analysis, canonical correlation, and multivariate regression. *Psychometrika, 46*, 139–142.

Nasser, F., & Takahashi, T. (2003). The effect of using item parcels on ad hoc goodness-of-fit indexes in confirmatory factor analysis: An example using Sarason's Reactions to Tests. *Applied Measurement in Education, 16*, 75–97.

Neyman, J. (1952). *Lectures and conferences on mathematical statistics and probability.* Washington, DC: U.S. Department of Agriculture.

Nunnally, J. C., & Bernstein, I. H. (1994). *Psychometric theory.* New York: McGraw-Hill.

Olsen, C. L. (1974). Comparative robustness of six tests in multivariate analysis of variance. *Journal of the American Statistical Association, 69*, 894–908.

Olsson, U. H., Foss, T., & Breivik, E. (2004). Two equivalent discrepancy functions for maximum likelihood estimation: Do their test statistics follow a non-central chi-square distribution under model misspecification? *Sociological Methods and Research, 32*, 453–500.

Pearson, K. (1896). Regression, heredity and panmixia. *Philosophical Transactions of the Royal Society, 187*, 253–267.

Pearson, K. (1908). On the generalized probable error in multiple normal correlation. *Biometrika, 6*, 59–68.

Pedhazur, E. (1982). *Multiple regression in behavioral research.* New York: Holt, Rinehart and Winston.

Penny, K. I. (1996). Appropriate critical values when testing for a single multivariate outlier by using the Mahalanobis distance. *Applied Statistics, 45*, 73–81.

Pierce, C. A., Block, R. A., & Agunis, H. (2004). Cautionary note on reporting eta-squared values from multifactor ANOVA designs. *Educational and Psychological Measurement, 64*(6), 916–924.

Pierre, M. R., & Mahalik J. R. (2005). Examining African self-consciousness and black racial identity as predictors of black men's psychological well-being. *Cultural Diversity & Ethnic Minority Psychology, 11*(1), 28–40.

Quintana, S. M., & Maxwell, S. E. (1999). Implications of recent developments in structural equation modeling for counseling psychology. *The Counseling Psychologist, 27*, 485–527.

Ramsey, J. B. (1969) Tests for specification errors in classical linear least squares regression analysis. *Journal of the Royal Statistical Society B, 31*(2), 350–371.

Rencher, A. C. (1998), *Multivariate statistical inference and applications.* New York: Wiley.

Rencher, A. C. (1992). Interpretation of canonical discriminant functions, canonical variates, and principal components. *The American Statistician, 46*(3), 217–225.

Reynolds, A. L., & Constantine, M. G. (2007). Cultural adjustment difficulties and career development of international college students. *Journal of Career Assessment, 15*(3), 338–350.

Rindfleisch, N., & Foulk, R. C. (1992). Factors that influence the occurrence and the seriousness of adverse incidents in residential facilities. *Journal of Social Service Research, 16*(3/4), 65–87.

Roy, J. (1958). Step-down procedure in multivariate analysis. *Annals of Mathematical statistics, 29*(4), 1177–1187.

Roy, S. N., & Bargmann, R. . (1958). Tests of multiple independence and the associated confidence bounds. *The Annals of Mathematical Statistics, 29*, 491–503.

Rubin, E. R. (2010). *The practice of social research.* Belmont, CA: Wadsworth.

Rubin, A. (2007). *Statistics for evidence-based practice and evaluation.* Belmont, CA: Thompson Brooks/Cole.

Rubin, D. B. (1987). *Multiple imputation for nonresponse in surveys.* New York: John Wiley & Sons.

Rubin, A., & Babbie, E. R. (2010). *Research methods for social work.* Belmont, CA: Thompson Brooks/Cole.

Satorra, A., & Bentler, P. (1994). Corrections to test statistics and standard errors in covariance structure analysis. In A. von Eye and C.C. Clogg (eds.), *Latent variable analysis: Applications to developmental research* (pp. 399–419). Newbury Park: Sage.

Schau, C., Stevens, J., Dauphinee, T. L., & Del Vecchio, A. (1995). The development and validation of the Survey of Attitudes toward Statistics. *Educational and Psychological Measurement, 55,* 868–875.

Scheffé, H. (1953). A method for judging all contrasts in the analysis of variance. *Biometrika, 40,* 87–104.

Schmidt, F. L., & Hunter, J. E. (1997). Eight common but false objections to the discontinuation of significance testing in the analysis of research data. In L. L. Harlow, S. A. Mulaik, & J. H. Steiger (Eds.), *What if there were no significance tests?* (pp. 37–64). Mahwah, NJ: Erlbaum.

Schuster, C. (1998). *Regression analysis for social sciences.* New York: Academic Press.

Seaman, M. A., Levin, J. R., & Serlin, R. C. (1991). New developments in pairwise multiple comparisons: Some powerful and practicable procedures. *Psychological Bulletin, 110,* 577–586.

Searle, S. (1987). *Linear models for unbalanced data.* New York: John Wiley & Sons, Inc.

Shapiro, S. S., & Wilk, M. B. (1965). An analysis of variance test for normality (complete samples). *Biometrika, 52,* 591–611.

Shevlin, M., Miles, J. N. V., & Bunting B. P. (1997). Summated rating scales: A Monte Carlo investigation of the effects of reliability and collinearity in regression models. *Personality and Individual Differences, 23,* 665–676.

Sokal, R. R., & Rohlf, F. J. (1995). *Biometry.* New York, NY: WH Freeman.

Small, N. J. H. (1980). Marginal skewness and kurtosis in testing multivariate normality. *Applied Statistics, 29,* 85–87.

Smith, S. P., & Jain, A. K. (1988). A test to determine the multivariate normality of a dataset. *IEEE Transactions on Pattern Analysis and Machine Intelligence, 10*(5), 757–761.

Smithson, M. (2003). *Confidence intervals.* Thousand Oaks, CA: Sage.

Sorbom, D. (1981). Structural equation models with structured means. In K. G. Joreskog & H. Wolds, (Eds.), *Systems under indirect observation: Causality, structure and prediction* (pp. 23–69). Amsterdam: North-Holland.

Srivistava, M. S. (1984). A measure of skewness and kurtosis and a graphical method for assessing multivariate normality. *Statistics and Probability Letters, 2,* 263–276.

Steiger, J. H. (1990). Structural model evaluation and modification: An interval estimation approach. *Multivariate Behavioral Research, 25,* 173–180.

Stevens, J. P. (1972). Four methods of analyzing between variation for the K-group MANOVA problem. *Multivariate Behavioral Research, 7*, 499–522.

Stevens, J. P. (1973). Step-down analysis and simultaneous confidence intervals in MANOVA. *Multivariate Behavioral Research, 8*(3), 391–402.

Stevens, J. P. (1980). Power of the multivariate analysis of variance tests. *Psychological Bulletin, 88*(3), 728–737.

Stevens, J. (1996). *Applied multivariate statistics for the social sciences.* New York: Routledge.

Stevens, J. (2002). *Applied multivariate statistics for the social sciences.* New York: Routledge.

Stevens, J. (2009). *Applied multivariate statistics for the social sciences.* New York: Routledge.

Stevens, S. S. (1946). On the theory of scales of measurement. *Science, 103*, 677–680.

Stevens, S.S. (1951). Mathematics, measurement and psychophysics. In S. S. Stevens (Ed.), *Handbook of experimental psychology* (pp. 1–49). New York: Wiley.

Student. (1908). The probable error of a mean. *Biometrika, 6*, 1–25.

Subbaiah, P., & Mudholkar, G. S. (1978). A comparison of two tests for the significance of a mean vector. *Journal of the American Statistical Association, 73*, 414–418.

Tabachnick, B. G., & Fidell, L. S. (2007). *Using multivariate statistics.* Boston: Allyn and Bacon.

Takane, Y., & Hwang, H. (2005). On a test of dimensionality in redundancy analysis. *Psychometrika, 70*(2), 271–281.

Tang, T., & Kim, J. K. (1999). The meaning of money among mental health workers: The endorsement of money ethic as related to organizational citizenship, behavior, job satisfaction, and commitment. *Public Personnel Management, 28*(1), 15–26.

Thompson, B. (1984). *Canonical correlation analysis: Use and interpretation.* Beverly Hills, CA: Sage.

Thompson, B. (1991). A primer on the logic and use on canonical correlation analysis. *Measurement and Evaluation in Counseling and Development, 24*(2), 80–95.

Timm, N. H. (1975). *Multivariate analysis with applications in education and psychology.* Monterey, California: Brookes-Cole.

Tintner, G. (1950). Some formal relations in multivariate analysis. *Journal of the Royal Statistical Society, Series B (Methodological), 12*, 95–101.

Tomarken, A. J., & Waller, N. G. (2005). Structural equation modeling: Strengths, limitations, and misconceptions. *Annual Review of Clinical Psychology, 1*, 31–65.

Toothacker, L. E. (1993). *Multiple comparisons procedures*. Thousand Oaks, CA: Sage Publications.

Tukey, J. W. (1953). *The problem of multiple comparisons*. Unpublished manuscript, Princeton University: Princeton, N. J.

van den Wollenberg, A. L. (1977). Redundancy analysis: An alternative for canonical analysis. *Psychometrika, 42,* 207–219.

Velleman, P. F., & Wilkinson, L. (1993). Nominal, ordinal, interval, and ratio typologies are misleading. *The American Statistician, 47*(1), 65–72.

Warner, R. M. (2008). *Applied statistics*. Los Angeles, CA: Sage Publications.

Widaman, K. F., & Thompson, J. S. (2003). On specifying the null model for incremental fit indexes in structural equation modeling. *Psychological Methods, 8,* 16–37.

Wilcox, R. R. (1987). New designs in analysis of variance. *Annual Review of Psychology, 38,* 29–60.

Wilks, S. S. (1932). Certain generalizations in the analysis of variance. *Biometrika, 24,* 471–494.

Winer, B. J. (1971). *Statistical principles in experimental designs*. New York: McGraw-Hill.

Wright, B. D. (1999). Fundamental measurement for psychology. In S. E. Embretson & S. I. Hershberger, (Eds.). *The new rules of measurement: What every psychologist and educator should know* (pp. 65–104). Mahwah, NJ: Erlbaum.

Yuan, K.-H. (2005). Fit indices versus test statistics. *Multivariate Behavioral Research, 40,* 115–148.

Yule, G. U. (1907). On the theory of correlation for any number of variables treated by a new system of notation. *Proceedings of the Royal Society, 79,* 182–193.

Index